Sustainable Building Standards and Guidelines for Mixed-Use Buildings

AuthorHouse™
1663 Liberty Drive
Bloomington, IN 47403
www.authorhouse.com
Phone: 1 (800) 839-8640

Published by AuthorHouse 09/28/2018

ISBN: 978-1-5462-5990-9 (sc)
ISBN: 978-1-5462-5989-3 (e)

Library of Congress Control Number: 2018910897

Print information available on the last page.

This book is printed on acid-free paper.

authorHOUSE®

Table of Contents

Section 1.0 Introduction 7

1.1 Purpose 8

1.2 Green Building Defined 9

1.3 Structure of Code of Practice 9

1.4 Design Philosophy 9

1.5 Project Strategies 9

1.6 Rating Systems 10

1.7 Corporate Policy 10

Section 2.0 The Design Process 12

2.1 General 13

2.2 Collaborative Design Engineering 14

2.3 Understanding Owner's Needs and Expectations 14

2.4 Measurable Green Criteria 14

2.5 Selecting a Designer 15

2.6 Building Information Modelling (BIM) 15

Section 3.0 Architectural Design and Planning 16

3.1 General 17

3.2 Design Process for Sustainable Architecture 17

3.3 Programming Phase 17

3.4 Schematic Design Phase 17

3.4.1 Design Development Phase 17

3.4.2 Code Compliance 18

3.4.3 Construction Documents Phase 18

3.4.4 Construction Phase 18

3.4.5 Commissioning Phase 18

3.5 Design Reviews 19

3.5.1 Internal Design Reviews 19

3.5.2 Owner Design Reviews 20

3.6 Managing the Design Process 20

3.6.1 Design Change and Modification Procedure 20

3.7 Ecological Assessment of Building Material 20

3.8 Indoor Air Quality (IAQ) 22

3.9 Storm Water Management 23

3.10 Water Efficient Landscaping 23

3.11 Design Tools 24

3.11.1 Energy Modelling 24

The Project Team 28

Section 4.0 Using the Sustainability Guidelines 30

4.1 General 31

4.2 General Requirements, Reviews and Approvals 32

4.3 Variances 32

Section 5.0 The Site 34

5.1 General 35

5.2 Microclimate 36

Sustainable Building Standards and Guidelines for Mixed-Use Buildings

RANJIT GUNEWARDANE

Section 6.0 Major Building Elements 37

6.1 General 38

6.2 Heat Island Effect 40

6.3 Biodiversity, Ecology & Landscaping 41

6.4 Transport 41

6.5 Building Envelope 41

6.5.1 Air Infiltration 41

6.5.2 Envelope Thermal Performance 41

6.6 U-Value 42

6.7 Glazing 42

6.8 Shading 43

6.9 Roofing 43

6.10 Noise Pollution 43

6.10.1 Using Sound Masking 45

6.10.2 Sound Control in Residential Apartments 45

Section 7.0 Energy and Buildings 47

7.1 Mechanical/HVAC Systems 48

7.2 Design Criteria 48

7.3 Comfort Conditions 48

7.4 Heating/Cooling Medium Production & Distribution 48

7.4.1 Absorption Chillers 49

7.4.2 Ground source Heat Pumps (GSHP) 50

7.4.3 Variable Refrigerant Flow (VRF) 50

7.4.4 Variable flow pumping 50

7.4.5 District Heating/Cooling systems 50

7.4.6 Sea Water Cooling 50

7.4.7 Combined Heat and Power (CHP) 51

7.5 Ventilation 51

7.5.1 Approaches 52

7.5.2 Kitchen Ventilation 52

7.5.3 Car Park Ventilation 53

7.6 Electrical System 53

7.6.1 Power/Electrical 53

7.7 Emergency standby power to dedicated supplies 55

7.8 Uninterruptible Power Supply (UPS) 56

7.9 Local Distribution/Protection 56

7.9.1 Distribution Panelboards 57

7.10 Harmonics 57

7.11 Metering Facilities 57

7.12 Motor Control Center (MCC) 57

7.13 Lighting 58

7.13.1 General 58

7.13.2 Control Systems 58

7.13.3 Daylight Harvesting 60

7.13.4 Exterior Lighting 61

7.13.5 Luminaires 62

7.13.6 Power over Ethernet (PoE) Lighting 63
7.14 Photovoltaics (PVs) 63
7.15 Electric Energy Storage Systems 63
7.15.1 Battery Energy Storage 64
7.16 Integrating Alternative Power Generation Systems 64
7.17 Vertical Transport & Escalators 65

Section 8.0 Energy and Building 66
8.1 Energy Conservation Measures 67
8.1.1 How to Achieve Building Energy Goals 67
8.1.2 Energy Management 67
8.1.3 Energy Audits 68
8.1.4 Energy Metering 70
8.2 Building Management System (BMS) 72
8.3 Thermal chilled-water storage 73
8.4 Renewable Energy 74
8.5 Miscellaneous Energy Saving Devices 74

Section 9.0 Water Management 76
9.1 General 77
9.2 Water Consumption 77
9.2.1 Assessing Facility Water Use 77
9.3 Water Conservation 78
9.3.1 Water Balance Modelling 78
9.4 Rain Water Harvesting 81
9.5 Landscape Irrigation 82
9.6 Water Tracking 83
9.6.1 Data retrieval and evaluation 83

Section 10.0 Waste Management 85
10.1 General 86
10.2 Definitions 86
10.3 Physical Characteristics 88
10.4 Methods of Waste Management 89
10.5 Construction Waste Management 92

Section 11.0 Products, Materials and Resources 96
11.1 Green Building Product Characteristics 97
11.2 Building Product Life Cycle 99
11.2.1 Three Phases of Building Materials 99
11.3 Specific Requirements 101
11.4 General Performance Requirements 101
11.5 Green Building Product Characteristics 101
11.5.1 Resource Efficiency 101
11.5.2 Low Emitting Materials 103
11.5.3 Affordability 104
11.6 Materials 104
11.6.1 Paints and Coatings 106
11.6.2 Coating System 107

11.6.3 Wood 107
11.6.4 Flooring Systems 108
11.6.5 Miscellaneous Building Elements 110
11.6.6 Adhesives, Sealants, and Finishes 112
11.6.7 Insulation 112
11.6.8 Regional Materials 113

Section 12.0 Green Information and Communication Technology 114

12.1 General 115
12.2 Internet Saves Energy and the environment 115
12.3 Information and Communication Technology (ICT) Impact on Pollution 116
12.4 Green computing 116
12.5 Ethics in ICT 116
12.6 Environmental Friendly ICT-Products 117
12.7 Sustainable Software Design 118
12.8 Systems Engineering for Designing Sustainable ICT- Based Architectures 119
12.8.1 Power over Ethernet (PoE) 120
12.8.2 Blown Fiber Infrastructure 120
12.9 Sustainable Cloud Computing 121
12.9.1 The Three-Ways to Cloud Computing 122
12.10 Green Audio Visual 123
12.10.1 Automated Power Systems 123
12.10.2 Business Conferencing 123
12.10.3 Remote Monitoring 124
12.10.4 Reuse AV products 124
12.10.5 Sustainable AV product purchasing 125

Section 13.0 Green Construction 126

13.1 What is Green Construction 127
13.2 Green Building Evaluation Systems 127
13.3 Green Construction Planning and Scheduling 128
13.4 Elements of Green Construction 128
13.4.1 Green Design-bid-build Project Delivery 128
13.4.2 Green Design-build Project Delivery 129
13.5 Material Conservation 130
13.5.1 Material Conservation Planning 130
13.5.2 Material Conservation Strategies 130
13.6 Products and Materials 131
13.6.1 Building Product Life Cycle 132
13.6.2 Three Phases of Building Materials 132
13.7 Site Layout and Use 133
13.8 Construction Waste Management 135
13.9 Material Storage and Protection 135
13.10 Providing a Healthy Work Environment 135
13.11 Construction Equipment Selection and Operation 136
13.12 Documenting Green Construction 137

Section 14. Green Project Certification and Closeout **138**
14.1 General 139
14.2 Commissioning and Testing 139
14.3 Building Commissioning Purpose and Objectives 140
14.4 The Design Intent Document 140
14.5 The Commissioning Process 141
14.6 Specifying Commissioning 142
14.7 Commissioning electrical systems 144
14.7.1 Thermographic surveys 145
14.7.2 Harmonic Analysis 145
14.7.3 Lighting Commissioning 146
14.7.4 Whole-Building Shutdown Tests 146
14.8 Commissioning HVAC Systems 146
14.9 Commissioning Non-HVAC Systems 147
14.10 Project Closeout 147
14.11 Post Construction Cleaning 148
Section 15.0 Facility Operations **149**
15.1 A Proactive approach for long-term success 150
15.2 Environment as a business issue 151
15.2.1 Organization Structure 152
15.2.2 Operations 152
15.3 Action Programs 156
15.3.1 Waste Management 156
15.3.2 Waste Audits 158
15.3.3 Product Purchase 158
15.3.4 Indoor Air Quality 163
15.3.5 External Air Emissions 165
15.3.6 Noise 168
15.3.7 Pesticides and Herbicides 170
15.3.8 Hazardous materials 171
15.3.9 Fuel storage 172
15.4 Operations and Maintenance 173
15.4.1 Artificial Intelligence (AI) and maintenance 174
15.4.2 AR and VR in Maintenance 176

Section 1.0

Introduction

1.1 Purpose

The purpose of the Sustainability & Environmental Standards and Design Guidelines is to represent the requirements established for design of a sustainable Mixed-Use Building.. This document contains scoping and technical requirements for Architectural and Engineering systems design. The intent is not to create a standard design but rather one of quality and consistency. It will establish a defined level of legislation, guest expectation and a means to measure and maintain a basic level of quality.

These requirements are to be applied during the design, construction, addition to, and alteration of sites, facilities, buildings, and elements to the extent required by regulations issued by local authorities. Standards are intended to be minimum requirements used by Architects, and other design professionals as a guide in the design process.

When specific design issues are not addressed, these Standards will be used as a guide to establish design intent in order to develop innovative design solutions which meet or, as governed by legislation, exceed the intent of the Standards.

These Design Guidelines will provide architectural and technical information to the design professionals for design of Green/Sustainable buildings. The design is one that will achieve high performance, over the full lifecycle, in the following areas:

- Minimizing natural resource consumption through more efficient utilization of non-renewable energy and other natural resources, land, water, and construction materials, including utilization of renewable energy resources to achieve net zero energy consumption.
- Minimizing emissions that negatively impact the global atmosphere and ultimately the indoor environment, especially those related to indoor air quality (IAQ), greenhouse gases, global warming, particulates, or acid rain.
- Minimizing discharge of solid waste and liquid effluents, including demolition and occupant waste, sewer, and storm water, and associated infrastructure required to accommodate removal.
- Minimizing negative impacts on the building site.
- Optimizing the quality of the indoor environment, air quality, thermal regime, illumination, acoustics/noise, and visual aspects to provide comfortable human psychological and physiological perceptions.
- Optimizing the integration of the new building project within the overall built and urban environment.

The guiding principles are drawn from published materials and standards developed by internationally recognised authorities including but not limited to:

- LEED – Leadership in Energy and Environmental Design produced by the US Green Building Council
- BREEAM – Building Research Establishment Environmental Assessment Method (UK)

- ASHRAE (American Society of Heating, Refrigeration and Air Conditioning Engineers)
- Estidama
- Green Globes 21

1.2 Green Building Defined

The term green building is defined in the ASTM Standard E2114-06a as building that provides the specified building performance requirements while minimising disturbance to and improving the function of local, regional, and global ecosystems both during and after its construction and specified service life.

1.3 Structure of Code of Practice

The Code of Practice is structured around the main stages in the building process. The sequence is more of a life-cycle than a linear process, and the structure of the code allows for a starting point at any stage.

The most important decisions affecting the impact of the building are taken at the earliest stages in building conception and design. The code encourages the involvement of building services engineers in early decision making.

1.4 Design Philosophy

Architectural design of the highest quality and being appropriate in their respective settings is essential to the success of the facility. It is the intent of the information provided in this manual to create a design that respect local building methods, new technologies and materials, as well as local cultural, religious and economic factors.

The "style" of the building should result from local indigenous architectural concepts being re-defined and re-invented. The resulting style should fit seamlessly and harmoniously into its setting, whether urban, suburban or rural in nature. It is essential that indigenous building methods that have qualities that are environmentally sensitive, and come from renewable source material, and allow for reduced energy consumption are used.

Designers should be required to create sustainable designs to international standards and to optimise the whole-life costs of facilities. The environmental impact of the materials and processes used in the construction of projects should be taken into account.

1.5 Project Strategies

The design process is the first crucial element in producing a green building. For design efficiency, the variety of owner's objectives and criteria, including sustainable/green goals, are described throughout this manual in order to minimize the potential of increased design costs.

In order to have a major impact on the performance [potential energy savings, water efficiency, maintenance costs, etc.] of a building, sustainability principles need to be applied at the very earliest stages of the design process.

For the design of a sustainable or green building project to be successful, it is important to understand the owner's goals. Possible goals could be a certain level of certification of a certain rating systems (e.g.,LEED Gold, BREEAM Excellent or Green Globes 3 Globes), a certain level of energy savings(e.g.,25% better than ANSI/ASHRAE/IES standard 90.1) , a net zero energy building (e.g., A+ according to ASHRAE's Building Energy Quotient (EQ) energy labelling program), or a sustainable building with system commissioning without official certification, from a rating system.

Of all the participants, it is the owner who is the most crucial when it comes to making a sustainable green building happen. Specific roles that the owner can in making a sustainable/green design successful include the following:

- Establishing the basic value system (i.e., what is important, what is not)
- Appointing a qualified and experienced commissioning agency
- Participating in selection of design team members
- Setting in schedules and budgets
- Participating in the design process, especially the early stages

The driver for sustainable green design is lowering the total cost of ownership in terms of construction costs, resource management and energy efficiency, and operational costs.

1.6 Rating Systems

Over the past decade, various green building rating systems or certification schemes were promoted across the globe. However, there are three most commonly used sustainability rating systems in the global coverage, they are LEED, BREEAM and GREEN GLOBES. LEED is the most recognized rating system mainly used in USA, Canada, India, Brazil and the Middle East. BREEAM is the BRE Environmental Assessment Method that mainly operated in UK and part of Europe including Netherlands, France, Spain, Germany, Sweden, Poland, Norway, Russia, etc. When comparing BREEAM and LEED there are similarities that sustainability issues are broken-down into a number of categories and assigned weightings, such as:

I. Management,
II. Energy,
III. Transport,
IV. Health and wellbeing,
V. Water,
VI. Materials,
VII. Land use
VIII. Ecology, pollution
IX. Sustainable sites, etc.

1.7 Corporate Policy

The corporate management recognises the importance of moral and ethical responsibilities in protecting the Environment, Health and Safety at work and wholly accept the aims and provisions of the Environment, Health and Safety Policy.

It is the aim of facility management to actively conserve natural reserves and energy, seek to provide healthy and safe working conditions and enlist the support of all management and staff in achieving these ends. The policies and guidance contained in the manual will be followed for fulfilling all legal requirements to create a pollution free environment and the protection of occupants, visitors, and staff of the building facility.

Managers based at this facility shall realise the importance of their responsibility for Environment, Health and Safety and must support the Facility Manager at all times on these issues by providing resources and time for employees to attend training.

As Facility Manager it is my intention to:

- Promote standards of safety, health and welfare which comply with corporate, national and local codes for employees to perform their work safely and efficiently and without risk to themselves and others.
- Maintain a safe and healthy workplace, safe systems and safe methods of work.
- To make available the necessary safety devices and personal protective equipment.
- Encourage full and effective consultation with staff on Environment, Health and Safety matters and to investigate any incidents reported to ensure that no recurrence takes place.
- Ensure that competent persons carry out the statutory assessments.
- Employees are reminded of their own legal and moral responsibility for conducting themselves in such a manner in their work so as not to expose themselves or others to risk. Employees must not promote or participate in pranks or practical jokes, which may result in an accident or injury.

This document will detail design criteria that would enable the following:

- Manage energy consumption, while maintaining optimum guest satisfaction and health and safety
- Efficiently manage and minimize waste production
- Pursue design elements benefiting the environment in the local community
- Utilize products and materials which have the least negative impact on the environment

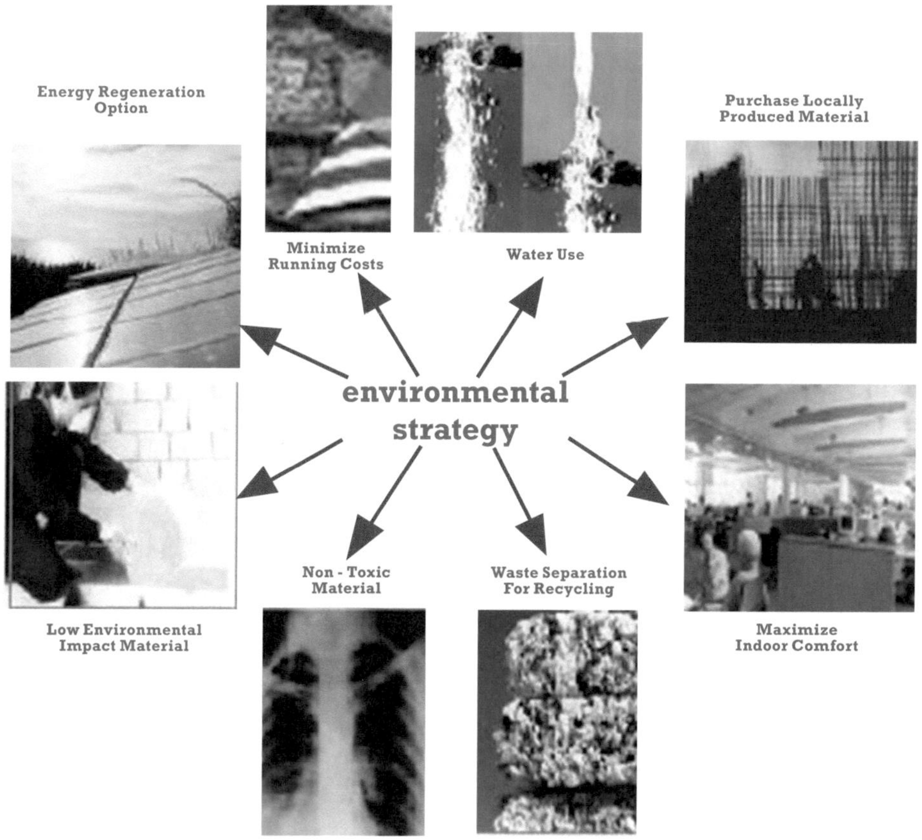

Section 2.0

The Design Process

2.1 General

The first task in a sustainable/green design project is forming the design team and the commissioning team. This team should include the design team leader [architect], the owner, the commissioning team leader, the design engineers, and building operations leader. A traditional project team includes the following members:

- Owner
- Project manager
- Architect
- HVAC & R engineer
- Plumbing/fire protection engineer
- Electrical engineer
- Lighting engineer
- Structural engineer
- Landscaping/site specialist
- Civil engineer
- Code consultant

An expanded project team for a sustainable/green design with commissioning would include the following members:

- Energy analyst
- Daylighting consultant
- Environmental design consultant
- Commissioning agent
- Construction manager/contractor
- Cost estimator
- Building operator
- Acoustical consultant

The preceding possible roles may be required on a reasonably large design project.

In addition to the standard tasks associated with the design project, the design team shall be responsible for developing and implementing new concepts that will create a green project. The addition of an experienced environmental design consultant [EDC] is highly recommended. The EDC's extensive knowledge will result in a cost-effective practical green design that meets the owner's requirements.

The project team — from the initial concept to the construction documents, construction, and building operations — must work as an integrated unit to achieve the goals set by the owner's objectives and criteria.

The design team's responsibility as part of the project team is to assist the owner with setting sustainabile green goals that include, but not limited to:

- Life-cycle cost optimization of energy-consuming systems, materials, and maintenance
- Minimize station of environmental impact
- Documenting basis of design
- Assisting with training of building operations and management staff during commissioning.

2.2 Collaborative Design Engineering

A key attribute of a well-designed, cost-effective green building is that it is designed in an integrated fashion, wherein all systems and building components work together to produce overall functionality and environmental performance. This has a major impact on the design process, for example determining the appropriate building envelope design and the use of modelling tools to design the building envelope. The building envelope shall be designed to adapt in a dynamic way to the shifting climate patterns.

The integrated design process shall include the following elements:

- Ensuring that as many off the design team members as possible are represented on the design team as early as possible
- Interdisciplinary work among architects, engineers, costing specialities, operations people, and other relevant persons engaged from the beginning of the design process
- Addition of an energy specialist to test various design assumptions through the use of energy and daylight simulations throughout the process, and to provide relatively objective information on a key aspect of performance
- Clear articulation of performance targets and strategies shall be updated throughout the process by the owner and the design team

2.3 Understanding Owner's Needs and Expectations

Unlike a traditional building project, with green building projects the owner's needs and expectations related to building operational criteria, life-cycle issues, an external commissioning agent working directly for the owner, or a specific certification of green building status by a third-party organisation must be thoroughly understood.

An owner's request for proposal (RFP) for the project that include the project criteria shall be developed. The owner's project criteria shall be a detailed and comprehensive package geared to the specific project. The owner's project criteria—sometimes referred to as either the design criteria package or statement of facility requirements—shall be part of the RFP that defines the owner's needs and expectations for the physical building project.

The design criteria package should define the specific green requirements that the designer is expected to meet, such as reduced site disturbance, optimized energy performance, and incorporation of daylighting and lighting control, among many other criteria. Alternatively, the owner's criteria could simply state that the design-builder achieve a certain green building status as certified by an outside third-party organization.

2.4 Measurable Green Criteria

The designer should ensure that the owner's green project criteria are measurable and achievable. The owner's RFP shall establish measurable green criteria that evolves into system performance criteria, and all the designer shall verify that the design criteria is complete, reasonable, and measurable. The designer shall confirm the measurable green criteria that will be the basis for both building design and commissioning.

The designer shall fully and completely define and understand the owner's measurable green criteria needs and expectations. What the owner asks for in its RFP may not be what it actually wants or needs. The designer shall identify any discrepancies between that explicitly stated in the RFP and implied during its analysis of owner needs. This step is very important in identifying areas where alternate systems, equipment, and materials will provide the owner with the same function at a reduced cost.

After reviewing the RFP and gaining an understanding of the owner's actual needs, the designer should develop measurable green criteria for the project using existing green building rating systems and industry codes, standards, and recommended practices. The design criteria shall be defined in terms of the project and documented for both approvals by the owner as well as use by the design team. These measurable green criteria will serve as the basis for design and later for testing to determine that the building meets the owner's stated requirements.

2.5 Selecting a Designer

The selection of a designer should be based on the project scope and the designer's experience and expertise. It is imperative that the designer has a background in green design and construction as well as thoroughly understands any green rating systems being used.

The designer performing the work or the principal of the design firm responsible for performing the design shall be a registered design professional.

In addition to professional licensing, designers shall be knowledgeable about green design and sustainable design strategies. Where appropriate, designers should demonstrate their green design expertise through a third-party certification or accreditation process. For example, the designer on a green building project using the U.S. Green Building Council's (USGBC) Leadership in Energy and Environmental Design (LEED) green building rating system should be a LEED-Accredited Professional (LEED AP), which involves successfully passing the LEED accreditation examination. In addition to demonstrating expertise, having a lead designer who is a LEED AP will streamline the green certification process for the building project and earn one point toward LEED accreditation.

2.6 Building Information Modelling (BIM)

Sustainable design is driving BIM solutions to ensure that buildings are designed, constructed, and operated in a manner that minimizes their environmental impact and are as close to self-sufficient as possible. Throughout the design process the building shall be a working digital prototype.

Multiple design options shall be developed and studied within a single model early in the design process, to not only see the building and provide conventional documentation for construction, but also to interact with other software to perform energy analysis and lighting studies.

Section 3.0

Architectural Design and Planning

3.1 General

The environment code of practice shall be structured around the main stages in the building process. The sections shall follow the sequence of a new building, from inception through to ultimate demolition and disposal. The sequence shall be a life cycle than a linear process.

Since Architects are the principal planners and designers of a construction project, they must communicate with the clients to ensure that they understand the project specifications, budget, and timelines. The client's environmental goal being construction of a luxury upmarket Mixed-Use Facility with a low carbon footprint, a firm or architect with demonstrated expertise in that area would be required.

Careful consideration is essential for the site selection, building orientation and form, the structure's envelope and the arrangement of spaces and zoning.

3.2 Design Process for Sustainable Architecture

On a green building project, the design process is usually divided into the following six phases:

- Programming
- Schematic design
- Design development
- Construction documents
- Construction administration
- Commissioning

3.3 Programming Phase

In the programming phase, the designer shall define the owner's needs and expectations. The designer shall translate these needs and expectations into measurable performance objectives that will serve as the basis for design. At the end of this stage, the designer shall review the programme and develop a conceptual cost estimate and schedule based on past experience with similar projects and industry information. This information should then be submitted to the owner for review and approval.

3.4 Schematic Design Phase

During the schematic design phase, the designer shall perform code reviews and any studies or testing required for the design. The designer will select materials, equipment, and systems that will be used in the design, and shall develop outline plans and specifications for the project in accordance with the program requirements. The designer shall perform a design review that includes a constructability review, value analyses, and life-cycle cost assessments as required. On completion of the design review, the designer shall update the cost estimate and schedule. The schematic design, along with the updated cost estimate and schedule, should then be submitted to the owner for review and approval.

3.4.1. Design Development Phase

In this phase, the designer shall prepare detailed plans and specifications that further define the project. The designer shall perform a design review that includes a constructability review, value analyses, and life-cycle cost assessments as required. After completing the design review, the designer shall update the cost estimate and schedule based on the increased design detail. The design, along with the updated cost estimate and schedule, shall then be submitted to the owner for review and approval.

3.4.2 Code Compliance

The latest version of the International Green Construction Code (IgCC), the full family of International Codes published by the International Code Council (ICC) and ADA Standards for Accessible Design should be applied to the design.

3.4.3 Construction Documents Phase

On traditional bid-build and construction manager projects, the designer shall complete the design in sufficient detail that it can be bid out and then built. On a design-build project, the detail that the designer provides on a traditional project may not be required because the work is not open for competitive bid, and both the designer and contractor are working together as a team. The level of detail required by the contractor will depend on the complexity of the project as well as other factors, such as the abilities of personnel in the field. The extent to which the contractor needs construction documents should be determined up front so the designer can budget and schedule for them if needed.

The designer should perform a design review at the completion of this phase before construction begins. Also, the cost estimate and schedule should be updated as they were in previous phases. The difference now is that the cost estimate will become the budget for the completion of the project, and the schedule will become the project's as planned schedule that will be used to track project progress. The design, along with the updated cost estimate and schedule, shall be submitted to the owner for review and approval.

3.4.4 Construction Phase

During the construction phase, the designer shall visit the site as required to determine that the work is being completed in accordance with the design documents. At the completion of this phase, the installation shall be inspected, put in service, and tested to ensure that it meets the owner's needs and expectations. During the construction phase, the owner monitors progress and, at the end of the project, inspects the work for compliance with the contract documents.

3.4.5 Commissioning Phase

The commissioning phase is mandatory to be included on green building projects that are required to be certified or verified as a green building using a third-party rating system. During the commissioning phase, the designer shall be required to assemble documents, create drawings, perform calculations, consolidate information from other members of the design team, and prepare the required submission for third-party certification or verification. This can be a significant effort and needs to be addressed in both the designer's schedule and budget.

The process in designing energy conserving architecture shall involve the following:

- Analyze applicable building codes and ensure that the project complies with them.
- Analyzing climatic conditions at the macro- and micro- site levels
- Identifying passive and active strategies applicable for the climate
- Design the building envelope, shape, and orientation

- to reduce energy loads while providing comfort
- Design structures that incorporate environmentally-friendly building practices or concepts, such as Leadership in Energy and Environmental Design (LEED) standards
- Integrate engineering elements into a unified design.
- Determine space layout, where rooms could be grouped into HVAC zones to reduce energy loads while providing desired comfort levels
- Perform predesign services, such as feasibility or environmental impact studies.
- Research case studies and building energy load data for similar building types in the same climate zone to identify typical resource consuming systems and benchmarked energy demands
- Apply computer-based simulation studies and related data sources to evaluate energy consumption against accepted metrics and benchmarks to demonstrate the performance-based sustainable outcomes for alternative designs

3.5 Design Reviews

The designer shall coordinate and schedule regular design reviews to ensure that the design is proceeding as planned and will meet the owner's requirements. These design reviews should occur at the end of each predefined milestone in the design process. These milestones may vary from project to project. It is essential that design reviews include project personnel and are scheduled early in the design process, so that design changes can be most easily and efficiently made at this time.

The project manager should schedule two types of design reviews:

- Internal design reviews
- Owner design reviews

3.5.1 Internal Design Reviews

Internal design reviews should be scheduled by the designer and performed at regular intervals as required by the complexity and scope of the project. The designer should be responsible for organizing, performing, and documenting the results of these reviews. Internal design reviews should include members of the design team, field personnel, suppliers and manufacturers, outside specialists, and others that are impacted by the design process. Internal design reviews should include not only a technical review of the design but also a review of the project schedule and budget.

A constructability review and value analysis should be performed as part of each internal design review as well. Constructability addresses the efficiency with which the system can be installed. Value analysis refers to, determining if the owner's needs and requirements can be met using alternative materials, equipment, and systems' at a lower cost. Constructability reviews and value analyses are most effective when conducted early in the design process.

On green design projects, each internal design review shall also include a review of the project's green goals to ensure that these goals are being met by the design. Where the owner's project criteria require that the building perform to a certain level or achieve a certain certification, the green design review would require that simulations be run with the updated design to ensure that it still meets the requirements.

Also, if a point system is being used to rate the building's sustainability, the criteria for awarding points or credits should be reviewed to ensure that the necessary number of points or credits would be earned by the building to meet the owner's project criteria.

3.5.2 Owner Design Reviews

Owner design reviews should be scheduled in accordance with the agreement between the owner and designer. Owner design reviews are essential for the designer to keep the owner informed of the design and its progress through these reviews. Owner design reviews should be scheduled only after corresponding internal design reviews are completed and should involve representatives throughout the owner's organization that are affected by the project.

3.6 Managing the Design Process

The detail in the design documentation will vary from project to project. The designer is required to determine the level of detail needed to obtain regulatory permits and efficiently construct the building. However, as the level of design detail increases, so does the design cost. Given this trade-off between installation efficiency and design cost, the designer needs to determine the level of design detail that is appropriate for the project.

3.6.1 Design Change and Modification Procedure

The designer shall establish and document design change and modification procedures with the owner at the commencement of the project. These procedures are important and should be a part of the designer's quality assurance program for projects. The owner's business is dynamic, and before the design is completed, changes in owner requirements may occur that necessitate changes to the system design. These changes must be documented and agreed upon by the owner's designated representative, because they might impact not only the technical performance of the system but also the project schedule and budget. In addition, the designer should be compensated for significant design changes.

Variation directive administration of the design require implementing systems and procedures to process potential changes.

3.7 Ecological Assessment of Building Material

In the selection of building materials, environmental impacts shall be considered in addition to the normal questions about cost, aesthetics, and durability. A primary consideration in material selection is to establish how the materials affect the building's energy performance.

The design team shall compute the baseline building performance rating according to the building performance rating method in Appendix G of ANSI/ASHRAE/IESNA Standard 90.1-2013 or USGBC approved method using a computer simulation model for the whole building project.

Additionally, the following attributes should be considered for building envelope design:

- The U-Value is a measure of how much heat passes through a given material. It is generally in W/m^2k, and shows the amount of heat lost in watts (W) per square meter of material when the temperature (k) outside is at least one degree lower. The lower the U-Value, the better the insulation provided by the material.

- The R-Value measures resistance to heat transfer through conduction and expresses insulation value of material, which is the inverse of the U-Value.
- The project team should specify the building envelope to achieve the lowest possible U-values as defined in this section and in any case to achieve 10% improvement in the proposed building performance compared to baseline building performance rating.

Building elements forming the external walls, roofs, and floors shall have an average thermal transmittance (U-Value), which does not exceed the following values:

Table 3.7.1 Average Thermal Transmittance Values

Building Component	Maximum U-Value W/m²C
External walls	0.25
Roofs	0.15
Floors	0.20
Glazed Outer Doors & Roof lights	1.30
Opaque Outer Doors & Hatches	0.60

Insulation, glazing, and mechanical/electrical systems shall be selected to enhance the buildings energy performance. Additionally, the material selection process shall identify

The life-cycle of building products

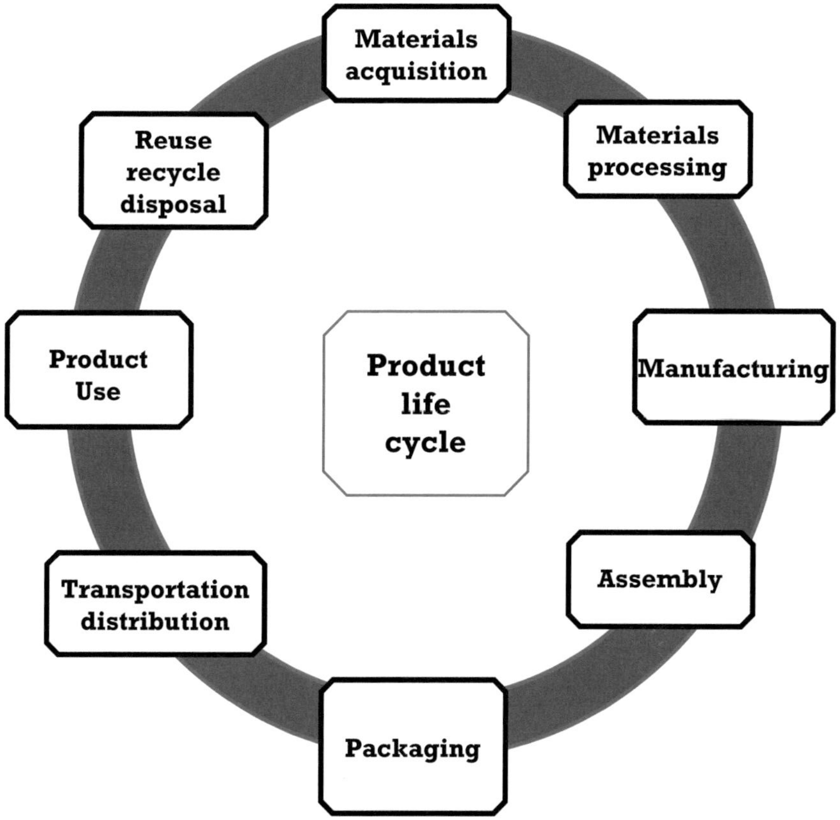

Figure 3.7

how the material will affect the health of building occupants, visitors and employees.

Other important environmental considerations for material selection shall include durability and maintenance. The amount of energy required to extract, process, transport of product shall be carefully reviewed. For example, in comparing the embodied energy office steel structure with the concrete one, it is clear that the production of steel is more energy intensive than concrete production. However, Steel contains high amounts of recycled steel content and the steel may be recycled indefinitely after the building is demolished.

Architects, interior designers, engineers shall ensure that products that incorporate recycled materials are fully incorporated in the building construction process. The value of the recycled content portion of a material or furnishing shall be determined by dividing the weight of the recycled content in the item by the total weight off all material in the item, then multiplying the resulting percentage by the total value of the item.

Materials and products that are extracted and manufactured within the region, thereby supporting the regional economy and reducing the environmental impacts resulting from transportation are preferred Natural materials produced from renewable sources that do not create hazardous byproducts in their manufacture shall also be referred.

The designers shall reduce the use and depletion of finite raw materials and long cycle renewable materials by replacing them with rapidly renewable materials. Consider the materials such as wool carpets linoleum flooring, cotton batt insulation etc.

Wood products that come from forest managed in accordance with the standards of the forest stewardship Council [FSC] shall be certified and specified by the architects and designers. The specifier shall clearly identify the use of wood and agri-fiber products that contain no added urea-formaldehyde dressings.

Reduce or completely eliminate the quantity of indoor air contaminants that are odorous, potentially irritating and/or half full to comfort and well-being of installers and building occupants. Specify Low-VOC materials, paints and coatings, carpet products and systems in construction documents. Ensure that VOC limits are clearly stated in each section where adhesives and sealants are addressed.

3.8 Indoor Air Quality (IAQ)

Among the requirements Buildings are expected to fulfill, indoor air quality (IAQ) is typically addressed through compliance with code requirements, which are based on industry consensus standards such as ANSI/ASHRAE Standard 62.1. IAQ affects occupant health, comfort, and productivity, and in some cases even building usability, all of which can have significant economic impacts for building owners and occupants.

Incorporating IAQ at the very beginning of conceptual design gets a number of key issues before the design team, enabling them to make informed decisions that will affect the project through the construction and occupancy phases.

The key elements of design for IAQ is as follows:

- Maintain proper building pressurization
- Select suitable materials, equipment, and assemblies for unavoidably wet areas

- Limit entry of outdoor contaminants
  - Control entry of Radon
  - Control intrusion of vapors from subsurface contaminants
- Design and build to exclude pests
- Control moisture and dirt in air-handling systems
- Control legionella in water systems
- Limit contaminants from indoor sources
  - Through appropriate material selection
  - Minimize impacts associated with cleaning and maintenance
  - Capture and exhaust contaminants from building equipment
- Maintain proper pressure relationships between spaces
- Provide appropriate outdoor air quantities for each room or zone
- Provide comfort conditions that enhance occupant satisfaction

3.9 Storm Water Management

Sites should be designed to treat storm water as a resource, capturing runoff and encouraging infiltration, and should use water efficiently. If existing imperviousness is less than or equal to 50%, a storm water management plan that prevents the post development 1.5 year, 24-hour peak discharge rate from exceeding the predevelopment 1.5 year, 24-hour peak discharge rate shall be designed. If the existing imperviousness is greater than 50%, a storm water management plan that results in a 25% decrease in the rate and quantity of storm water runoff shall be designed.

The options for controlling storm water runoff are numerous and are often implemented in combination to create systems appropriate for site and project budget.Permeable interlocking concrete pavements (PICPs) are recognized as an effective means of storm water control.

Permeable systems - whether concrete, natural, or a combination thereof - reduce or eliminate rainwater flowingoffsite and instead capture, filter, and direct it back to the ground. Permeable paving shall be specified to minimize impervious surfaces, which would contribute to several LEED credits.

A scheme to reuse storm water volumes denigrated for non-portable uses such as landscape irrigation, toilet and urinal flushing and custodial uses shall be incorporated in the design. A Rainwater Harvesting System consisting of an assembly that collects, stores, and distributes rain water for use in situ; including water treatment as appropriate to intended use shall be provided. A below-ground storage tank system shall be considered.

3.10 Water Efficient Landscaping

Selection of appropriate plant materials for landscaping will require the use of native species, as these plants will have the best chance for survival with the minimum amount of water and maintenance and will support the native wildlife indigenous to the site.

The planting scheme should grade down in height from taller trees at the back, to shrubs in the middle, to a strip off taller grass at the front.

Landscape irrigation systems shall minimize water waste and maximize system effectiveness and savings by incorporating technology advances elements into their irrigation management activities.

Irrigation system design shall incorporate the newest generation of smart controllers for irrigation systems that connect to the internet, so system operators can access them anytime from anywhere. The benefit of these controllers is that they do not need to be located near the main controller.

The controllers shall have capability to allow users to operate systems more efficiently by setting specific programs or having pre-set programs run based on soil and weather conditions. The controllers shall be capable of accounting for such weather elements as wind, rain, temperature, and humidity, which keeps turf and other vegetation healthy and minimizes water use.

The smart controllers shall have pre-set programs based on soil profiles, which allows the systems to monitor the rate at which water leaches and absorbs. As a result, the system will only water the grounds when the soil needs hydration. The irrigation system shall incorporate moisture sensors.

Flow sensors shall be incorporated into a facility's irrigation system to assist the system operators determine the amount of water the system uses regularly, as well as the rate of pumping. These sensors that control the amount of water used will identify potential problems in the system. For example, the sensors will signal the controller to automatically shut down the flow if an overflow or leak occurs.

Flow sensors shall be programmed to assist in gathering essential data to make more informed decisions about landscapes' irrigation needs.

The use of gray-water is an irrigation option that could tremendously assist with water conservation. Although Gray-water is not safe to drink because it comes from sinks, showers, tubs, and washing machines, and it has not gone through water processing, it has many positive uses for landscape irrigation. Gray-water contains valuable nutrients that are beneficial to grass, plants, and flowers. From an economic and environmental standpoint, reusing gray-water accomplishes many goals, including reducing a facility's water bill and conserving the drinkable water supply.

Since gray-water should never be stored for more than 24 hours because it starts to smell, designers should avoid pumps and filters, and have a three-way valve installed to switch between the gray-water system and the sewer and septic system.

3.11 Design Tools

Design tools such as the ones described below help maximise the project's potential and assure that the design will function as intended.

3.11.1 Energy Modelling

The complex interaction of building systems, exterior climate, and occupant actions makes understanding building energy performance a design challenge for which energy modeling is particularly well suited. The building components and characteristics that determine energy use by the building need to be included in an energy model simulation. Every decision made as a building design progresses from the conceptual stage through the completion of construction documents has potential impact on energy use. Waiting until the later stages of the design process limits the impact and benefit that energy modeling can provide. Understanding how much energy a building will use and how it compares to the maximum energy use allowed by an energy code or to a baseline building starts with conceptual development.

Energy modelling is the discipline that models the energy flows in buildings and between a building and its (local) environment, with the aim of studying the heat and mass flow within buildings and their (sub)systems under given functional requirements that the building must satisfy. Models should be computational in nature, and implemented in the form of a computer simulation that replicates a part of physical reality in the machine.

The design team shall begin initiating computer simulations early in the design process for maximum effectiveness. A detailed load analysis through computer simulation can identify energy-saving opportunities early in the design process. With a baseline energy model created at the outset of the project, the energy performance should be monitored throughout the design process. Changes in the design should be entered into the model to assess the energy impact.

The computer energy simulation provides a method to test the integration of various design solutions to verify that they are meeting design goals. Decisions about building form, materials, and systems should be tested and adjusted to improve performance.

The methodology used should be based on the principles of integrated resource planning. and shall adopt a "bottom-up" approach which essentially looks at the opportunities to reduce power demand by improving energy efficiency

Energy modelling techniques shall be used by the consultants to simulate the interaction of building systems and how they affect overall energy performance with the aim to:

I. Optimise energy balance

II. Reduce cooling and heating energy needs

a) Energy modelling software includes Energy-Plus, eQuest or equal and approved

b) BIM (Building Information Modelling)

BIM is a combination of software and methodology that translates building geometry from a design model into an interactive digital model that can be used by architects, engineers and contractors to optimise

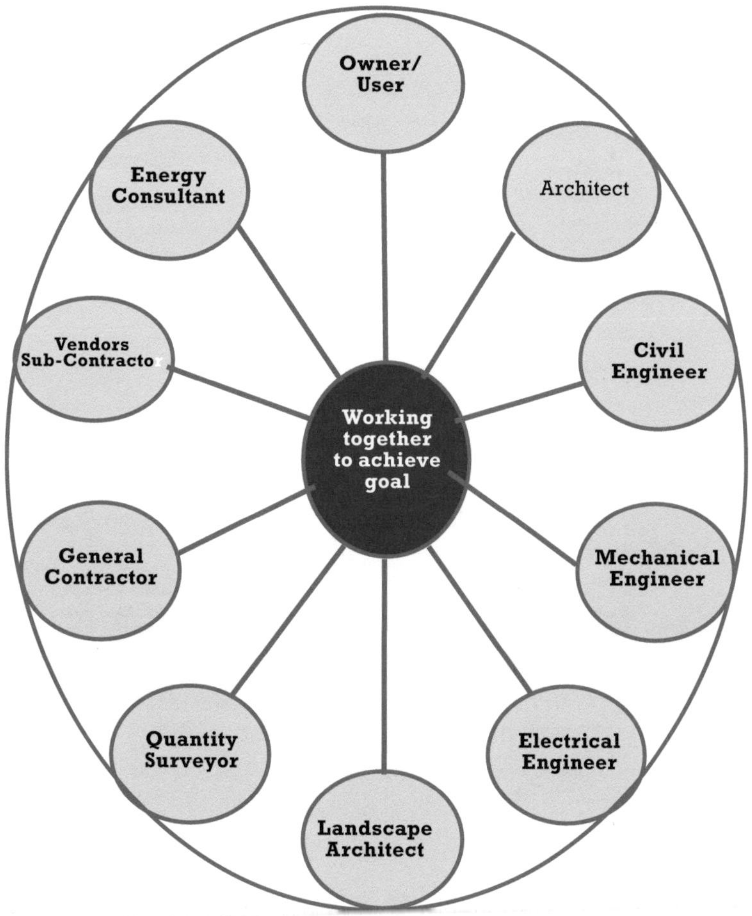

the design and construction process, construction sequencing and building operational energy use.

c) Daylight Modelling & Lighting

Daylighting shall be properly integrated with electrical lighting control systems. Lighting analysis and daylight modelling simulates lighting and daylighting into the building and the interactions of the building envelope and energy use with the aim to:

- Maximise daylight use
- Reduce direct electrical energy consumption for electrical lighting

The lighting designer shall influence some of the building's glazing properties, automated shade controls and other features to improve the benefits of daylight harvesting. Whilst solar heat gain coefficient (SHGC) is an important factor, visual light transmission (VLT) for a window must be addressed for daylight and view, and their benefits. The focus should be on quantifying the contribution of daylight harvesting and glare management. The design team shall consider external shading devices or internal window controls such as blinds to allow occupants to make adjustments to ensure comfort. For skylights or roof monitors, they would also consider diffusing glazing, baffles or louvers to diffuse sunlight.

The design team shall evaluate the impact of changes in glazing performance, assess the effect of potential glare sources such as white or bright reflective surfaces on occupants and help define any repositioning of occupant's activities needed to avoid potential sources of glare.

In order to successfully implement "daylight harvesting" the building should have automated controls that either turn off or dim artificial lighting in response to the available daylight in the space. Factors such as bi-level and multi-level switching or dimming capability as well as separate circuiting of luminaires in daylighted zones shall be considered. The control systems should always be supplemented with manual override to accommodate individual differences.

Automated systems should include optical sensors (photocells) that read ambient light levels to both maintain a base level of illumination, by using as much free natural daylight as possible, and occupancy sensors to shut lights off when spaces are unoccupied.

Autodesk tools like Revit, Ecotect and 3dsMax as well as plugins like Radiance and Dayism are the main software programs and tools recommended for lighting and daylight simulation.

Planning/ Concepts	Schematic Design	Detailed Design	Construction/ C_x	Post Occupancy
Energy target testing and budget alignment	Equipment selection support based on building part-load profiles	Value engineering support	Product submittal evaluation	Measurement and verification
Assessment of programmatic impacts	Lifecycle cost analysis support	Control sequence optimization	Control sequence comparison	Facility management optimization
Preliminary utility cost ranges	Occupant thermal comfort evaluation	Product specification support	Utility incentive documentation	
		Goal check-ins		
		Utility incentive estimation		

Table 3.10 Energy model use opportunities

The Project Team

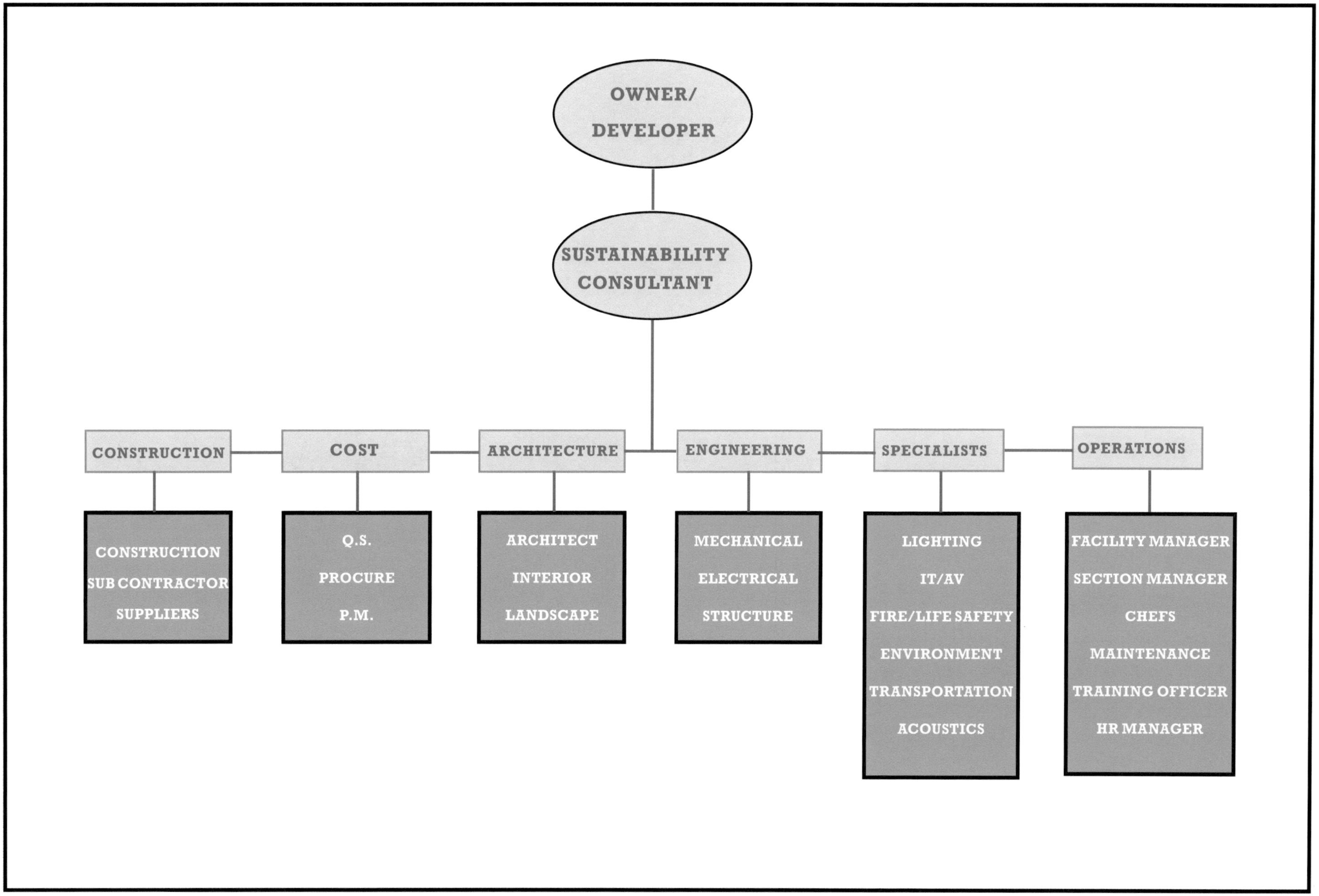
OWNER/
DEVELOPER
SUSTAINABILITY
CONSULTANT
CONSTRUCTION
COST
ARCHITECTURE
ENGINEERING
SPECIALISTS
OPERATIONS
CONSTRUCTION
SUB CONTRACTOR
SUPPLIERS
Q.S.
PROCURE
P.M.
ARCHITECT
INTERIOR
LANDSCAPE
MECHANICAL
ELECTRICAL
STRUCTURE
LIGHTING
IT/AV
FIRE/LIFE SAFETY
ENVIRONMENT
TRANSPORTATION
ACOUSTICS
FACILITY MANAGER
SECTION MANAGER
CHEFS
MAINTENANCE
TRAINING OFFICER
HR MANAGER

Section 4.0

Using the Sustainability Guidelines

4.1 General

These Sustainable Building Standards and Design Guidelines (SBSDG) will assist architects and other design professionals in developing imaginative, efficient and marketable design solutions as well as ensure a safe and environmentally sound structure, while developing plans for new Mixed-Use Building construction, renovation of existing Mixed-Use Buildings and conversions. Sustainable buildings aim to limit their impact on the environment mainly through energy and resource efficiency. Sustainable Mixed-Use Building development is expected to bring together at least the following three key elements:

a) Environmental sustainability

b) Economic sustainability

c) Social Sustainability

Sustainable development was first defined by the Brundtland Commission, a subsidiary body of the United Nations in 1987 as:

"Sustainable development is development that meets the needs of the present without compromising the ability of future generations to meet their own needs"

This idea has been extended for enterprises to develop sustainable strategies in the Triple Bottom Line (TBL) model (see Figure 4.1).

The SBSDG requires that higher standards be met beyond local codes. Therefore, it is important for the owner to engage the services of competent design professionals who are familiar with these requirements and how they affect Mixed-Use Building design. This will serve to speed up the design as well as approval process of local governing authorities. It will also save time and money through the development of creative solutions which comply with legislation.

a) The designers shall be required to adopt Sustainable Design Criteria in all new build and renovation projects.

b) These guidelines will assist architects, engineers and other design and management consultants in developing an integrated building design with a number of innovative ideas on energy optimisation and sustainable practices as well as creating a comfortable environment for Mixed-Use Building occupants and visitors.

c) Sustainability Criteria shall be applied from the beginning of a project, to ensure that the Client's needs and objectives are coordinated and the design is tested and budgeted.

d) Owner's representative shall review the designs regularly to ensure that the sustainability issues are properly addressed in the Project Brief and at sign-off stages.

e) Owner to prepare a detailed plan for continual improvement in the environmental performance of their Mixed-Use Buildings. The plan shall include goals with benchmarks for all their Mixed-Use Buildings to:

   I. Reduce waste generation by 10% by 2020

   II. To register the Mixed-Use Building properties for achieving LEED or equivalent certification

   III. For its properties to use renewable energy for at least 20% of its needs

4.2 General Requirements, Reviews and Approvals

Design is a creative activity by which client's needs and objectives are collected, interpreted and expressed in three-dimensional physical solutions.

Design management is an important activity in the design process as it involves the coordination, analysis and testing of the design, as well as the management of the different stakeholders involved.

To ensure that optimal design, value for money and buildability are achieved, due care and attention need to be given to the proper management and coordination of all design activities throughout the design process. In particular, the interaction between the different design disciplines requires a well-coordinated teamwork structure. Design management encompasses all of the coordination, analysis and design testing activities that a project requires. For effective design management and coordination, it is necessary to appoint a manager with appropriate management skills to ensure the design process operates efficiently. Such a person is usually the design team leader.

To ensure its effectiveness, the Project Coordinator should draw up a programme which includes the main areas of activity (i.e. Planning, Implementation and Review) up to project occupation. In relation to the design development activity (part of the Planning Stage) the Design Team Leader should, as soon as the other principal consultants are appointed, draw up details of design responsibilities and milestones for each consultant and illustrate them on a project programme which should be part of the Project Execution Plan.

Design should be a staged process during which a number of approvals / sign-offs are required from the client. Approvals are usually given as part of the formal project review structures. In each case, there should be sufficient information for the client to give informed approval. The timing and sequencing of client approvals may differ from project to project, depending on how the design process is carried out.

Various regulations and laws apply to the design of projects, both building and civil engineering, and to their owners and users. Projects should be designed so that approvals from all relevant Statutory Authorities can be obtained.

The owner's representative should review designs regularly to ensure that they satisfy the needs expressed in the Definitive Project Brief and Sign off on designs.

4.3 Variances

The purpose of these Design Guidelines is to establish a level and consistency of quality in design and construction of Mixed-Use Buildings. However, due to variations in building construction, site conditions and other variables, some aspects of these guidelines may not be possible to achieve. When specific requirements cannot be met, alternative solutions should be explored which achieve the same or similar level of quality and meet the intent of the guidelines.

When considering variances, a distinction should be made between new construction and the renovation or conversion of existing structures.

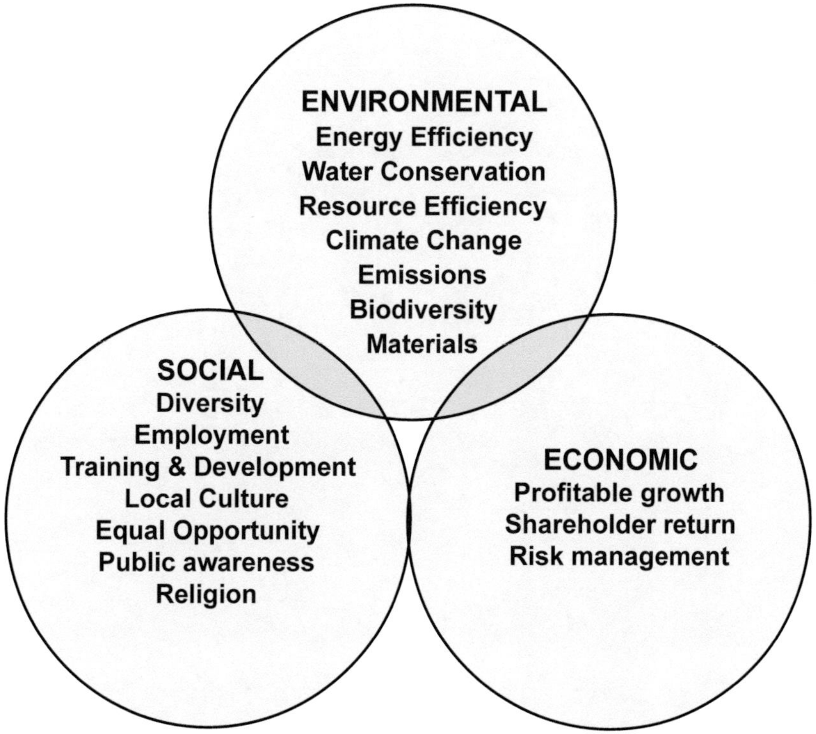
ENVIRONMENTAL
Energy Efficiency
Water Conservation
Resource Efficiency
Climate Change
Emissions
Biodiversity
Materials
SOCIAL
Diversity
Employment
Training & Development
Local Culture
Equal Opportunity
Public awareness
Religion
ECONOMIC
Profitable growth
Shareholder return
Risk management

Figure 4.1

Section 5.0

The Site

5.1 General

Developed land use has tremendously increased over the years. This trend is replicated throughout much of the world as populations grow and migrate to urban areas of the world.

With this growth, land that could have been put to other uses is lost. These uses could include land for growing food, and diversified ecosystems. As paved areas grow, rainwater is no longer attenuated at the point of contact with the ground and immediately runs off taking contaminants and eroding soil Waterways get polluted.

Developed land tends to consist of a high percentage of hardscape - such as parking areas and building roofs. These are heated by the sun during the day and re-emit that heat at night which can raise local air temperatures. This can increase air conditioning loads for a building and also impact insects and other wildlife.

A site should be chosen, designed, and constructed, to minimize the impacts listed above. The social impact of the site should also be considered along with the ease of access to those populating the site.

Global climate change models anticipate a broad range of impacts. Many of these changes and impacts have direct implications for the development of land. If increase in frequency of precipitation is anticipated, both in terms of rainfall and its distribution, increased erosion and slope destabilization can be expected. As sea levels continue to rise, increase in shore and beach erosion should be anticipated along coastlines. Rising sea levels will also complicate floods on tidal influenced rivers and streams.

Site planners and designers will have to respond to these trends in both retrofitting existing facilities and designing new projects.

Sustainable site planning should include considerations of the impact of development on the local ecosystem, global ecosystem, and the future.

During the site selection process, give preference to those sites that do not include sensitive site elements and restrictive land types. Select a suitable building location and design the building with the minimal footprint to minimize site disruption. Strategies shall include stacking the building program, tuck under parking etc.

Site-specific characteristics such as sun angles, wind exposure, and availability of infrastructure services shall be documented. Field surveys and desk studies shall cover topography and geology of the site and soil conditions

Design shade constructed surfaces on the site with landscape features and minimize the overall building footprint. Consider installing high-albedo and vegetated roofs to reduce heat absorption, and control heat island effect.

Additionally, the site shall be investigated to identify if it has any features of value to wildlife, such as mature trees, hedgerows, ponds and areas of meadow. Such features shall be integrated into the building development.

A site analysis should be performed to note the area's size and the amount of storm water that must be acomodated Many storm water best management practices have unique pros and cons and is site dependent.

Construct ponds that are large as possible, with a simple

shape and gently sloping agers. The pond shall be made to accommodate shallow water, with a deep area to maintain water and oxygen levels in dry periods. The pond shall be located away from trees and shrubs.

Table 5.1 Environmental Issues in Site Selection	
1.	Identify low environmental impact services
2.	Assess external microclimate (seasonal temperatures, rainfall patterns, shade, prevailing wind)
3.	Analyze air and water quality
4.	Identify local noise factors
5.	Analyze transport implications
6.	Establish available infrastructure and local amenities
7.	Investigate waste disposal and recycling opportunities
8.	Use of low ecology/derelict land
9.	Establish the water table

The development shall be channelled to urban areas with existing infrastructure, and preserve habitat and natural resources.

Transportation survey shall be performed to identify transportation needs of the Mixed-Use Building occupants, visitorsm and staff. The building may be cited near mass transit.

During the site survey identify site elements that could be adopted into the master plan for development of the project site. Clearly marked construction boundaries shall be established to minimize disturbance of the existing site and restore previously degraded areas to their natural state.

The Consulting Team shall

a) Analyse site-specific characteristics including sun angles, wind exposure, access to transportation hubs, and availability of infrastructure services.

b) Carry out filed surveys and desk studies to cover topography and geology of the site and soil conditions

c) Carry out an assessment of natural systems and the impact the site will have on the surrounding eco-systems and prepare a Natural Systems Assessment report highlighting different natural systems and areas to be protected and allocated for habitat creation.

d) Prepare a detailed Environmental Impact assessment (EIA) shall be provided at the start of the project

e) Prepare a Construction Environmental Impact Management Plan (CEIMP) for approval by the local Environmental Department.

5.2 Microclimate

The Consulting Team shall:

a) Evaluate local site microclimate including winds and shaded areas

Identify passive design features that can be incorporated in the design to improve thermal comfort and reduce energy requirements

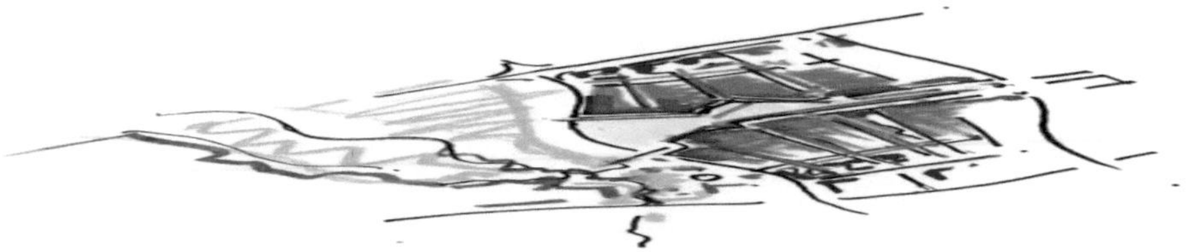

Section 6.0

Major Building Elements

6.1 General

The location of the building on the site is a critical element of site planning. The building should be located to minimize the impact on the site while maximizing the function and quality of the designed space. The location should be selected through a combination of managing the solar influences of the site, balancing the earthwork, and maximizing the utility and aesthetics of the site.

The building's form, orientation and envelope construction shall be determined by considering appropriate passive design strategies. Passive design refers to the creation of building elements and configurations that take advantage of the physical environment of the site, such as climate data, building site's existing or potential topography with its landscape details, and potential optimization of natural on site resources. Anticipated warming in most places will result in increasing cooling costs, and properly locating a building on the site and planning plantings to lower energy costs shall become an important element of the site plan.

The project team shall use design simulation software for sun path diagram and shading model analysis to optimize the geometry of the building. During the process of establishing the orientation and form of the building, the following need to be addressed:

- Effective use of daylight
- Satisfactory building heat balance
- Utilization off seasonal solar gain and avoiding glare
- Minimum exposure to prevailing wind and wetting
- Minimum extent of noise penetration using appropriate areas as buffer zones
- A minimum summertime temperature without the need for mechanical cooling
- The provision of a satisfactory visual environment with good personal control
- Efficient use of materials taking into account energy and without compromising adaptability or lifespan
- Appropriate use of thermal capacity

In predominantly cold climates, the compact cube shape with a tight, well insulated envelope is optimal to reduce heat loss from conduction and infiltration. However, elongation in an East-West axis will increase exposure to the winter sun, thus increasing winter heat gain. Reducing overall glazing area, specifying high performing windows and placing the bulk of Windows on the equator facing exposure while minimizing glazing on other orientations shall be considered.

For a predominantly hot climate, consider creating large shaded surface areas in contact with the outside. This can be accomplished with the creation of open courtyards, deep wraparound porches and/or multiple building wings with several openings for ventilation paths and aligned with prevailing wind patterns.

The building envelope encompasses the entire exterior surface of the building, including roofs, walls, foundations and floor slabs that enclose the interior space and provides the thermal barrier between the indoor and outdoor environment.

A building façade does many things:

- Provides thermal control
- Limits air flow
- Delivers daylighting
- Prevents water entry

The type of external cladding material used on the building has a big impact on water absorption. Even the best, and most carefully installed cladding systems must manage water that penetrates the exterior skin.

Depending on the climate zone, the most appropriate of the following three approaches shall be selected to keep wind-driven rain out of walls:

- Barrier systems
- Drainable assemblies
- Pressure-equalized rain-screens

The designer shall provide detailed description of method to be adopted to stop the water at a single plane, the joints between cladding members must be resistant to the penetration of driven rain over time. Project located in not so severely cold or wet climate, a barrier system may be a very practical solution. Where the project is located in climate that has severe weather conditions, a "perfect" air barrier and a "perfect" vapor barrier is required.

The design of thermal mass and weight of building materials internally and externally will be a function of climate. Buildings with high internal gains and large diurinal temperature swings may require high mass and heat absorbing materials.

Windows and glazing systems shall be carefully selected to ensure the safety and comfort of building occupants, in addition to maintaining character of the building façade. Consequently, the window design and glazing selection are critical to the performance and appearance of the building envelope. Spectrally selective products applicable to glazing technology that allow for higher transmittance of visible light than that of the heat contributing infrared portion of the spectrum shall be provided.

External shading devices shall be an integral part of the building envelope design to limit solar heat gain and glare. Horizontal shading devices to be placed above windows on south facing walls, and vertical louvres/shading devices would be effective for east, west and north facing windows.

Use continuous insulation where possible on walls and roofs, and account for the thermal bridging of window framing and metal or wood studs in walls. The minimum R-value required for the climate zone for the building envelope shall be established using ASHRAE Standard 90.1 – 2007 (International Climate Zones) as a baseline for minimum insulation R-values. The tables outline R-value standards for roof – insulation above deck, Walls Above Grade, Walls Below Grade, floors and Slab-on-Grade Floors shall be followed.

Green walls and roofs can assist in lowering the building's cooling loads by adding to the overall insulating performance of the building envelope as it creates a better thermal buffer. The International Building Code (IBC) requires all buildings to be designed in accordance with the International Energy Conservation Code (IECC),

The green roof systems addressed shall be typified by a top layer of living plant material and soil (growth media or engineered soil) supported on the roofing assembly below.

If the vegetation is deciduous, then it could be designed in a way to work harmoniously with the building's solar shading needs over the seasons.Green roofs provide wildlife habitat and attract birds.

The building orientation should be optimised so that its position allows maximum daylight while minimising unwanted solar gains. The placement, design and the selection of fenestration materials are extremely important for a high-performance building.

The Consulting Team shall:

a) Use daylight and energy computer modelling techniques to optimise solar orientation balances the penetration of daylight and reduction of cooling load.

b) Maximise extent of south and north facing building elevations, minimising extent of east and west facing building elevations.

c) Shading & Vegetation

d) The Consultant Team shall use design simulation software for sun path diagram and shading model analysis to optimise the geometry of the building

e) External shading devices shall be an integral part of the building envelope design to limit solar heat gain and glare

f) All pedestrian links to the facility must be shaded using materials with a designer selected Solar Reflectance index (SRI).

g) Horizontal shading devices to be placed above windows on south facing walls

h) Vertical louvres/shading devices are effective for east, west and north facing window

6.2 Heat Island Effect

Urban Heat Island (UHI) Effect arises due to absorption of incident radiation from the sun by built concrete surfaces of buildings and roads and then releasing it in the form of heat. Some of the negative impacts of increase of temperature due to UHI are:

- Increase in energy consumption;
- increase in emissions of air pollutants and GHG's;
- increase demand on water and adversely affect human comfort due to elevated day and night temperature

To reduce the Urban Heat island Effect the following guidelines shall be followed:

a) Use of reflective materials with high reflective index (SRI) values for roofs

b) Minimise hardscape areas

c) Use permeable materials and surfaces for driveways, access roads and parking spaces such as vegetated roofs, porous pavement and grid pavers

d) Externally painted walls must have a minimum Light Reflective Value of forty-five percent (45%)

e) Sloping roofs should have an SRI greater than 78

f) Heat rejection and exhaust equipment must be installed not less than 3m above ground level of the building

6.3 Biodiversity, Ecology & Landscaping

a) Integrate biodiversity considerations into the site planning decisions for continued conservation of eco-systems and for the long-term success of the facility.

b) Maximise open spaces and use the green spaces for recreation and for reducing heat island effect

6.4 Transport

To reduce the environmental impact of traffic to and from the facility the following measures shall be included in the design and operation of the Mixed-Use Building:

a) To efficiently manage the number of delivery calls to the facility

b) Reduce the number of daily waste collection calls

c) Offer guests the option to arrive and depart using 'group transport'

d) Use electric powered, hybrid or flex-fuel vehicles for facility shuttle and service use

e) Provide a secured bicycle storage area for staff. Provide bicycle storage spaces for 5% of the full-time employee count.

Make provision for electric charging points in the car park for electric cars/scooters

6.5 Building Envelope

The Design Team shall ensure that appropriate thermal comfort levels are achieved through thermal modelling and selection of building envelope design and controls to maintain a thermally comfortable environment for all occupants within the building.

6.5.1 Air Infiltration

Indoor Air Quality is one of the most important characteristics of green buildings. The design team should ensure that an airtight building envelope and airtight are fully described.

Since the pressure at any point on a building envelope depends on building form and local topography. It should be calculated for a given wind speed and direction by using the pressure coefficient (Cp). The pressure coefficient shall only be derived from wind tunnel tests.

Computational Fluid Dynamics (CFD) or models, such as the British Standard ventilation prediction model or the CIBSE single-sided model shall be employed to predict the whole building ventilation or infiltration rate.

Air infiltration of untreated outside air should be prevented by pressurising the building with filtered and treated outside air. The building pressurisation control shall be managed by the building management system.

6.5.2 Envelope Thermal Performance

The materials that comprise the structure and façade of the building form the thermal envelope of the building, the main barrier to external heat and solar radiation.

a) The consultant shall calculate the baseline building performance rating according to the building performance rating method in Appendix G of ANSI/ASHRAE/IESNA Standard 90.1-2007 or USGBC approved method using a computer simulation model for the whole building project.

a) In general, the following attributes should be considered for building envelope design:

6.6 U-Value

The U-Value is a measure of how much heat passes through a given material. It is generally in W/m^2k and shows the amount of heat lost in watts (W) per square metre of material when the temperature (k) outside is at least one degree lower. The lower the U-Value, the better the insulation provided by the material.

The R-Value measures resistance to heat transfer through conduction and expresses insulation value of material, which is the inverse of the U-Value.

a) The project team should specify the building envelope to achieve the lowest possible U-values as defined in this section and in any case to achieve 10% improvement in the proposed building performance compared to baseline building performance rating.

b) Building elements forming the external walls, roofs, and floors shall have an average thermal transmittance (U-Value), which does not exceed the values given in Table 3.7.1.

c) Vapour barriers to be provided to restrict ingress of moisture through the building walls, floors and roof given the high humidity in coastal areas.

d) The consultant is required to take into account thermal bridging in the thermal calculations and modelling. A thermal bridge occurs as a result of a physical connection across a cavity through the insulation and results in increased heat transfer across the bridge. The junctions that need to be accounted for include wall-floor junctions, wall-roof junctions, lintels, jambs, cills, intermediate floors, balconies, corners, party walls and other significant junctions. It is expressed in terms of linear thermal transmittance values, and it should be evaluated using thermal simulation software, following agreed conventions and standard.

e) For Hot and dry climates, materials with high thermal mass to be provided to give the building adequate thermal mass to absorb solar heat during the day and release the heat at night. For hot-humid climates, materials with low thermal mass to be used.

6.7 Glazing

Glazing design is important for the overall energy use because windows bring light and heat into occupied spaces. Optimising glazing materials and window design can have a significant impact on building cooling load and thermal performance.

a) The consultant shall specify glazing with low-E coatings to reduce Solar Heat Gain Coefficient (SHGC). The low-E coating is to be applied to the inner pane of a double pane window to be effective. Other ways to reduce cooling load include the application of reflective coatings, tinted glass, fritted glass and exterior shading.

b) SHGC between .20-.35 is recommended. The SHGC measures how well a translucent or transparent enclosure product blocks heat caused by the sun. It is a fraction of external solar radiation that is admitted through a window both directly transmitted and absorbed and subsequently released inwards.

c) The SHGC is expressed as a percentage in decimals

d) between 0 and .87. SHGC replaces the shading coefficient (SC) as the standard indicator, which was expressed as a number between 0 and 1. The relationship between SHGC and SC is:
SHGC = SC x .87

e) The SHGC needs to be established in conjunction with and under consideration of the Visible Light Transmittance (VLT). **Visible light transmittance** is an optical property that indicates the amount of visible light transmitted. Most VLT values are between 0.3 and 0.8. The higher the VLT, the higher the amount of light is transmitted. A high VLT is desirable to maximize daylight.

f) The consultant shall consider the orientation and size of the windows. Window to wall ratios should be optimised for daylighting, insulating properties and cost.

6.8 Shading

Shading devices reduce solar gain whilst allowing for natural lighting.

a) Shading devices shall be located externally and designed following sun path analysis for the particular location. Consider light-colored surfaces on shading devices such as overhangs, louvers, or light shelves. These light surfaces will bounce diffuse sunlight into the building

b) The consultant shall use design simulation software for sun path diagrams and shading model analysis to optimise the location, size and geometry of shading devices.

c) Structural materials and elements used for external shading shall have a minimum reflectance Index of 29.

Consider a deep exterior wall section that can be used to self-shade the window surfaces with overhangs and vertical fins.

6.9 Roofing

a) Roofs shall be designed with high solar reflectivity and high emissivity, typically while colour. The benefits of cool roof are:

- Reduced peak cooling load
- Reduced roof surface temperature and extended useful life
- Reduce heat island effect

b) Roofing material must have a SRI of 29 or higher for steep slope (>2:12) roofing and a SRI value of 78 or higher for low slope roofing.

Solar reflectivity or reflectance is the ability of a material to reflect solar energy from its surface back into the atmosphere. The SR value is a number from 0 to 1.0. A value of 0 indicates that the material absorbs all solar energy and a value of 1.0 indicates total reflectance.

The **Solar Reflectance Index** is calculated according to ASTME 1980 using values for reflectance and emissivity. Emissivity is a material's ability to release absorbed energy.

6.10 Noise Pollution

Different buildings demand different acoustical conditions within, The design team shall take into consideration the importance of controlling appropriately numerous sources contributing to the exterior ambient noise of mixed-use buildings, due to the blend of commercial, cultural, residential and industrial uses.

Environmental noise is a measure of sound levels immediately "outside the boundary fence". Standards for environmental noise are based on the premise that there are tolerable levels of noise that can be defined, which do not present a hazard to human health or harm the environment.

Noise levels are measured in decibels (dB), and the noise standards listed in this guideline refer to the 'A' frequency rating, dB(A), which covers the range audible to the human ear.

'Leq' is a measurement unit applied to an average number of decibels over a specified period.

The guidelines describe minimum requirements for managing environmental noise and vibration that can potentially either be hazardous to human health or harm the environment.

Sources of noise and vibration that have the potential to impact on facility occupants, visitors, employees and local communities shall be:

- Identified
- Analysed
- Quantified (either by direct measurement or using appropriate methods of estimation)

The International Building Code (IBC) section 1207, International residential Code (IRC) section AK 102 (Air Borne Sound), and IRC section AK 103 (Structural Borne sound) shall apply to common interior walls, partitions, and floor/ceiling assemblies between adjacent residential apartments and adjacent public areas such as corridors, stairs or service areas.

The table below provides details of maximum internal noise levels for both external noise sources and building services noise.

External Noise Intrusion		
Location	**External Noise**	**Services Noise**
Public spaces	40 dB L_{Aeq}	NR 35
Residential Apartment – night-time	27 dB L_{Aeq} 45 dB L_{Amax}	NR 22
Residential Apartment – daytime	32 dB L_{Aeq}	NR 25
Meeting Facilities – daytime	35 dB L_{Aeq} 50 dB L_{Amax}	NR 30
Spa and Fitness Centre – daytime	40 dB L_{Aeq}	NR 40

The design team shall take the following into consideration during the design stage:

- Noise from external sources shall be measured over a 30-minute period (note that, for either daytime or night-time, the maximum noise level must be met during any 30-minute period). Noise measurements of external noise break-in shall be carried out 2m from the external wall and 1.5m above floor level
- The Mixed-Use Building layout shall avoid positioning of health clubs and spas (including

- swimming pools) above noise sensitive spaces, unless all due measures are employed to control transfer of structure-borne sound to below
- When kitchens, laundries and stores *must* be located above sensitive areas impact sound level must be improved through the installation of floating concrete/screed floors. The specialist acoustic consultant's specifications are to be followed
- In new construction locate noisy equipment as far from noise-sensitive guest occupied areas as possible
- Control mechanical noise in the environment by enclosing the equipment, designing an acoustical barrier, or including sound attenuators. Where fully enclosing mechanical equipment is not possible, because some equipment needs access to outside air in order to function properly, and where air inlet and outlets are necessary, a sound attenuator, acoustical louver, or splitter shall be used
- Install an acoustically absorptive surface on the wall facing the mechanical equipment, whenever there is another structure opposite. The surface should have a minimum Noise Reduction Coefficient (NRC) of 0.90
- Sound insulation between vertically and horizontally adjacent spaces, including crosstalk via ductwork and service risers, must achieve the minimum levels of performance set out for Sound Insulation Matrices, by the local authorities
- Acoustically absorbent finishes shall be installed within common areas from which direct access is gained to residential apartments, with suitable conditions being provided by fitting standard carpet and pad
- All equipment (whether located in equipment rooms or occupied spaces), elevator cabins, elevator motors and ductwork/pipe work systems shall be isolated from the building structure in order to ensure that vibration within the floor of any occupied room is controlled to specialist acoustic consultant's specifications
- All air intakes and exhaust points shall be located at least 20m away from residential apartment windows

6.10.1 Using Sound Masking

The design team should ensure providing building occupants with improved speech privacy, noise control and acoustic comfort. Using sound masking as the starting point for interior planning the design team should set the base level of background sound throughout the facility.

Sound masking technology should be engineered so their output can be tuned post occupancy in order to provide a spectrum specifically designed to balance acoustic control and comfort. The localized computer tuning - where the software adjusts the system's output in order to meet the masking curve throughout the treated area - will mean a minimum background sound level is areadily deliverable component of architectural acoustic design. The design team should use this known controlled level as the foundation for the remainder of their acoustical plan.

6.10.2 Sound Control in Residential Apartments

The International Building Code (IBC) requires good floor/ ceiling acoustics in multifamily construction. Two of the principal measurement standards for acoustics in multifamily construction are:

- sound transmission class (STC), which pertains to the amount of airborne sound contained by a given building element (i.e. walls, doors, windows, and floor/ceilings); and
- impact insulation class (IIC), which deals with impact noise (i.e. footfall, chair scrapes, and dropped objects) transmitted through a floor/ceiling system.

The International Code Council (ICC) created ICC G2-2010, Guideline for Acoustics.which provides two levels of acoustical performance: 'acceptable' and 'preferred.' Both exceed code minimums for airborne and structure-borne noise.

These levels give a clear direction on what levels should be targeted for desired acoustical performance, depending on the building type. As the names suggest, when one wants a building that has an acceptable level of acoustical separation, 'acceptable' is targeted. When one is designing a building on the higher end of market rate or luxury level, or has tenants or owners sensitive to noise, the desire should be for a 'preferred' level of performance.

The design team must determine the level of acoustic performance to which to design based on the clients expectations. This should not be a matter of just meeting code, rather must be approached with answers to the following questions:

I. When considering the amenities offered to tenants, how important are the acoustics of the apartment? In other words, how important is the quality of life related to acoustical privacy?

II. Does the design team want to just meet code or do they want 'preferred' performance to meet client expectation and meet the expectations of the luxury apartment overall quality expectations.

III. Once the level is determined, which method makes the most sense for achieving that performance level?

It is important that design teams continue to be educated on new products and adapt their specifications to ensure they meet defined levels of sound control that tie directly to the end user's satisfaction with their home and living space.

Section 7.0

Energy and Buildings

7.1 Mechanical/HVAC Systems

Mechanical services and controls have a large impact on the energy performance of a building, typically as much as the fabric. The design and selection of mechanical systems and equipment are dependent upon several factors including:

a) Climate.

b) Orientation of building.

c) Building envelope design.

d) Heating/Cooling and ventilation loads.

e) Floor to floor heights.

f) Occupancy.

g) Occupancy type e.g. residential apartments, public areas, swimming pools, ballroom, back of house areas.

h) Efficiency of equipment selected.

i) Availability of equipment.

j) Space required for mechanical plant.

k) Maintenance requirements.

l) Life Cycle of mechanical plant and equipment selected.

7.2 Design Criteria

Design calculations shall be based on procedures and data published by ASHRAE or CIBSE and local Meteorological Department. Outdoor design conditions should be the 1% cooling dry-bulb design point for the specific geographic location of the building. The indoor design conditions should be based on the facility brand standards.

The consultant shall use diversity factors on the overall heating/ cooling load including occupancy, equipment and lighting loads. Diversity factors shall be as described in current version of ASHRAE Handbook- Fundamentals.

7.3 Comfort Conditions

Thermal modelling shall be undertaken to demonstrate that appropriate thermal comfort levels are achieved for casual visitors, guests and staff in accordance with CIBSE Guide A (CIBSE AM11) and ASHRAE Handbook.

7.4 Heating/Cooling Medium Production & Distribution

The goal of these systems is to provide the Mixed-Use Building facility with a reliable source of heating and cooling to provide the required output energy using the minimal amount of input energy.

A sound sustainable cooling strategy should consider the following five steps:

a) Reduction and modulation of heat transfer;

b) Use of direct and indirect ventilation processes;

c) Heating/Cooling energy from renewable sources;

d) Analysis of free cooling options;

e) Implementation of sustainable distribution systems.

The key issues that need to be considered as part of developing a successful sustainable Heating/ Cooling design shall include the following:

- Sustainable heating/cooling energy must not be considered an independent part of the building but needs to be

- integrated within the building design.
- The process of designing sustainable heating/cooling systems is essentially *iterative* and *progressive*; this requires close collaboration between the architects and building service engineers, and ideally should take into account the views of a number of stakeholders, including Mixed-Use Building owners and facility manager

Where active mechanical cooling is required in a building, this can be provided by a variety of systems giving varying efficiencies. These are mechanical central chillers, using one of:

- Absorption chillers; or
- Vapor compression refrigerant chillers.

Conventional heating/cooling plant produces hot or chilled water that circulates around the building supplying fan coil units or other emitters where necessary. Efficiencies vary between systems using different types of heating/cooling media.

Chillers can 'dump' heat from the building to the external environment via:

- Wet cooling towers;
- Dry cooling towers;

To maximize efficiency of heating hydronic systems, the consultant team may select the right-size hydronic boilers for peak and low-load conditions and look at the overall impact the boiler design and control strategy have on the system. The designers may consider adding one or more boilers to include redundancy and load matching and provide the correct system turndown. By staging a multi-boiler system, individual condensing boilers can be set to operate at low fire, thereby making better use of the inverse efficiency characteristics.

7.4.1 Absorption Chillers

Once the design team attribute the electric load for conventional cooling equipment as a thermal load using absorption chillers, the Mixed-Use Building could have a good balance between electric and thermal load all year round. Absorption chillers could be weaved into the central mechanical plant operation in many ways, four of the primary ways are as follows:

a) Part of Combined Cooling, Heat and Power (CCHF)or tri generation application

b) As a standalone gas fired absorption chiller application

c) Using renewable solar as the heat source for the refrigeration cycle

d) Waste heat application

Adding absorption chillers to the central plant opens up the possibility to install natural gas emergency generators used for everyday power and use the waste heat from the generators for cooling during the summer and hot water generation in the winter.

The gas fired absorption chiller would be a good consideration in areas where the electricity costs are high and natural gas costs are more stable. Where a dependable solar energy is available, the solar absorption chiller may be considered.

Where absorption chillers are incorporated into a central plant, the design team should ensure that the equipment manufacturers and the engineering integrators jointly commission the central plant as one system.

7.4.2 Ground source Heat Pumps (GSHP)

GSHP systems extract heat from the ground in winter months and reject heat to the ground in summer months. They use smaller amounts of electricity than air source heat pumps

As a failsafe approach, in cold climates 15% - 20% ± biodegradable propylene glycol antifreeze should be added to the circulating water to allow flow at 22 deg.F to 18 deg. F and contaminate the groundwater due to a pipe breakage.

7.4.3 Variable Refrigerant Flow (VRF)

VRF systems offer an alternate HVAC solution. Its key attributes include zonal control, energy efficiency, and indoor air quality (IAQ).

Zoning of the system is paramount to capturing the possible energy savings. In the heat pump mode, all indoor units connected to the system could either operate in heating or cooling mode at any given time, with each unit having control over its zone temperature.

The ventilation-air strategy of the system should be given careful consideration to ensure the best IAQ. The following three main methods of introducing outside air into the system:

- Direct to the unit
- Via a dedicated outside air system (DOAS) or energy-recovery ventilator (ERV)
- Directly to the space.

The design team shall apply the Air Conditioning, Heating and Refrigeration Institute (AHRI) Standard 1230 establishes definitions, classifications, test, rating and data requirements, as well as nameplate and marking requirements.

7.4.4 Variable flow pumping

Variable flow pumping at minimum static pressure should be included as a principal system optimization strategy.

An efficient variable flow water pumping system should actively compensate for system changes when operating at capacities less than design load.

7.4.5 District Heating/Cooling systems

District heating is considered the long-term solution. When all the local district heating networks share the same operation conditions, it will, in the long run, be possible to connect them to a large combined heating and power plant.

In a district cooling system (DCS) the cooling source is generally equipped with electrical compressor driven chillers which generate chilled water to cope with the cooling demand. Chilled water is distributed by variable speed pumps which creating a pressure differential between the supply and return pipes of the distribution system. The interface between the DCS and the building air-conditioning system is the substation.

7.4.6 Sea Water Cooling

Depending on the geographic location of the Mixed-Use Building and local natural resources, seawater could be used to cool the condenser of central cooling plants producing chilled water. seawater cooling has the important advantage of slow seawater temperature fluctuation; in the summer months, the temperature is generally below the ambient air temperature.

The utility of the cold-water delivery system can be improved by diverting excess flow potential at night to storage tanks (space permitting), then allowing the stored cold-water to augment daytime flow. Alternatively, the excess night flow could be used as a heat-sink for an ice-making process, where the ice is used during the day to reduce the chilled water temperature (increase the temperature differential).

7.4.7 Combined Heat and Power (CHP)

When designed to operate independently from the grid (in "island" mode) CHP systems can provide critical power reliability while providing electric and thermal energy on a continuous basis, resulting in daily operational cost savings. CHP systems can be configured in a number of ways to meet the specific reliability needs and risk profiles, and to offset the capital cost investment for traditional backup power measures.

While using CHP to take advantage of natural gas resources, businesses are using CHP to deliver energy savings and increased reliability from the growing availability of renewable resources such as biogas.CHP technology has enabled the availability of packaged CHP systems.These systems are engineered and assembled offsite, with heat exchangers, electronics, and controls assembled in a complete package. CHP systems can facilitate the integration of renewable technologies like wind and solar.

7.5 Ventilation

All enclosed spaces in a building shall be ventilated in accordance with the requirements of this section, current version of the IBC and local codes where applicable. Every space in the building shall be designed to have outdoor air ventilation as follows:

a) Each space shall be ventilated with a mechanical system capable of providing an outdoor air rate no less than the larger of:

- The conditioned floor area of the space times the applicable ventilation rate from the current version of ASHRAE 62.1 Standard or equivalent standard.If local codes require more ventilation than specified in the standard, the local code requirements must be met.
- 15 cfm per person times the expected number of occupants.
- For spaces without fixed seating, the expected number of occupants shall be either the expected number specified by the building designer or one half of the maximum occupant load assumed for egress purposes, whichever is greater. For spaces with fixed seating, the expected number of occupants shall be determined in accordance with the latest version of ASHRAE Standard 62.1.

During short-term episodes of poor outdoor air quality, ventilation can be temporarily decreased using a short-term conditions procedure from the current version of ASHRAE Standard 62.1. Similarly, consideration may be given to increasing outdoor air ventilation rates beyond those required in the standard where the quality of the outdoor air is high, and the energy consumed in conditioning it is not excessive.

7.5.1 Approaches

Conditioning and transporting ventilation air accounts for a significant fraction of building energy use. The Strategies presented below may be considered in reducing the energy required to deliver good IAQ.

a) Use Dedicated Outdoor Air Systems (DOAS) where appropriate so that conditioned 100% outdoor air is delivered directly to occupied spaces or to other heating/cooling units that serve those spaces. DOASs would make it easier to verify that the required amount of outdoor air is delivered and could reduce the total outdoor air required relative to other systems. DOASs may be combined with energy recovery or Demand Controlled Ventilation (DCV) to further reduce energy use.

b) Energy recovery ventilation (ERV) reduces energy use by transferring energy from the exhaust airstream to the outdoor airstream. Use ERV where appropriate when energy recovery ventilation is required by energy standards and when it can have favorable economics as well as it can improve humidity control and reduce the risk of mold growth.

   In general, there are two types of energy recovery ventilation devices: total ERVs that transfer heat and moisture between incoming and exhaust air and heat recovery ventilators (HRVs) that do not transfor moisture. Types of energy recovery systems include an energy recovery wheel and fixed plate with latent transfer, fixed plate, heat pipe, and runaround loop systems.

   In humid climates, properly ventilating Mixed-Use Building spaces and also controlling indoor humidity is essential. ASHRAE Standard 62.1 requires that relative humidity in a space be limited to less than 65% during cooling conditions. Using an energy recovery system with latent transfer capability can provide improved humidity control while saving energy

c) DCV will vary ventilation airflow based on measures of the number of occupant's present. It could be particularly cost-effective for spaces with intermittent or highly variable occupancy. The benefit of DCV increases with the level of density, transiency, and cost of energy. Occupancy categories such as theaters, auditoriums/public assembly spaces, gyms, restaurants, office conference rooms, etc. would be the most appropriate for DCV. Certain aspects of DCV controls maybe beneficial office buildings in ensuring that the design ventilation rates are supplied under all operating conditions. For example, continuous measurement of outdoor airflow rates and indoor CO_2 levels could assist facility management personnel find ventilation system faults or make adjustments to the HVAC system setpoints, thus avoiding overventilation or underventilation relative to the design or code requirements

d) Natural ventilation could be a low-energy strategy that provides a pleasant environment in mild climates with good outdoor air quality. Mixed mode ventilation can provide similar benefits in additional climates through the limited use of mechanical equipment. Use Natural or Mixed-Mode Ventilation Where Appropriate.

7.5.2 Kitchen Ventilation

In general, local authorities may require the use of either the

International Mechanical Code (IMC) or the National Fire Protection Association (NFPA) Standard # 96 or a combination of both. Codes and Standards will dictate the performance effectiveness of the exhaust of the kitchen using a variety of hood designs. On the air supply side, it is essential to take into consideration the comfort of the kitchen occupants, air flow patterns so that they would not have an impact on the hood performance or cool plated food awaiting collection by restaurant service staff, and air balance in the kitchen space under varying exhaust flow conditions.

Variable Air Volume (VAV) exhaust control strategies with reliable controls and system components should be considered.

7.5.3 Car Park Ventilation

Enclosed or underground car parks require ventilation systems to prevent the build-up of carbon monoxide during normal day to day use of the car park and assist firefighting operations. Relevant building regulations and codes shall be applied to the design using options for mechanical ventilation.

If the car park is above ground, the possibility of using natural ventilation compared to mechanical ventilation shall be fully investigated.

7.6 Electrical System

A **Green** or **Sustainable** building is designed:

- To save energy and resources, recycle materials and minimise the emission of toxic substances throughout its life cycle,
- To harmonise with the local climate, traditions, culture and the surrounding environment, and
- To be able to sustain and improve the quality of human life while maintaining the capacity of the ecosystem at the local and global levels

The design team shall utilize products and services that are integrated, intelligent and networked solutions that allow building occupants to use electricity in complete safety, use automation everywhere, improve energy efficiency and ensure a high-quality power supply and manage building utilities and communication networks.

As described in the introduction to Energy Modelling, the methodology used for developing the electrical distribution system should be based on the principles of integrated resource planning and adopt a "bottom-up" approach which essentially looks at the opportunities to reduce power demand by improving energy efficiency.

7.6.1 Power/Electrical

a) Design

With the ever-increasing efficiencies of electrical system products, energy losses from these products have decreased. This efficiency has a secondary benefit of reducing the cooling required due to the associated reduction in heat load.

The electrical service installations should be designed to serve all areas of the Facility and provide for high availability in the 7x24 environment. This requires integrated systems that enable interconnection, data collection and analysis, and need minimal or no human intervention, thus enabling faster decision making.

The reliance on information systems that operate 24 hours per day, 7 days per week, requires the system to have one infrastructure to manage all the buildings systems and can work across all trades: mechanical, electrical and Information Technology (IT). By tagging all transactions, the Building Facility Manager should be able to segment devices and collect much needed data. Being IT enabled the facility manager should have access directly to a device with an IP address.

The design team should apply the Internet of Things (IoT) principles and processes that allow devices and locations to generate and share data with each other and with other IT systems via the internet. Artificial Intelligence (AI) shall be leveraged to do autonomous control, enhance security and interact with occupants of the building.

b) Transformers

The most efficient way to transport electrical power from point A to point B is to use the highest voltage available. Higher voltage equates to lower currents for a constant amount of power. The concept of using the highest voltage available applies to power distribution within Mixed-Use buildings.

A transformer is required at each point along the distribution path, wherever a change in voltage is necessary.

The design team and Mixed-Use Building owner/operator strive for high performance, net zero buildings, maximizing transformer efficiencies. The transformer continues to consume energy 24 hours per day, 7 days a week. Since transformers sized for maximum peak demand using conservative diversification factors allowed by the National Electrical Code (NEC) never experience 100% load, transformers designed for maximum efficiency at higher loading will operate inefficiently.

The design team should apply current US Department of Energy (DOE) 10 CFR 431 standard to all low voltage dry type transformers. If the Mixed-Use Building load is known to produce harmonics, K-rated transformers or harmonic mitigating transformers should be considered.

Transformer K - factor Ratings	
K - factor	Load Type
K-1	Linear loading
K - 4	Solid-statetate electronics
K - 9	Medium - density solid-state electronics
K - 13	Heavy-density solid-state electronics
K - 20	Switching loads and variable frequency drives
K - 40	High-order harmonics and switching loads

Since no-load losses of transformers vary with temperature rise ratings, the design team should select the best choice for the particular application based on the calculated average loading on the transformer.

Branch circuit conductors that run from the final circuit breaker to the outlet or load shall be designed to allow a 3% maximum voltage.

Provide a transformer paralleling system to allow redundancy, so that the building shall be able to continue to operate without interruption should there be a loss of one transformer. This design follows the N+1 policy for capacity. However, should there be a loss of two or more utility transformers, the building's standby generators should start and parallel, while the MV switchgear disconnects from the utility system. The building will then remain on the generator source until the utility source is restored, at which time the generators will parallel with the recovered source and, once the utility voltage has stabilized, reconnect the building load to it without interruption.

c) A reliable electrical supply

The electrical supply should have a reliability of - 99.98%. If a supply of this reliability is not available, then an additional electrical supply system should also be provided in the form of an independent electrical generating plant, or a totally independent power supply from an alternative generating source. An availability in excess of 99.98% is required (i.e. not more than one-hour interruption per year) to obviate the need for the generator.

The incoming supply, generator supply and electrical distribution should be arranged to prevent a total loss of power in the event of failure of a single item.

To ensure good quality power supply, the designer shall consider the effects of harmonics on electrical systems. Since electronic devices are nonlinear loads, they create both voltage and current distortions. The sustainable approach would be to implement a system-wide solution with harmonic mitigating transformers.

d) Energy Efficiency

ASHRAE 90.1 standard requirements shall be applied in the design for energy efficiency. The concept shall ensure reduction of losses in the electricity transmission and distribution system. The voltage drop shall not exceed 2% for feeders and 3% for branch circuits.

- All the transformers shall meet the requirements for high efficiency transformers as outlined in ASHRAE 90.1 Table 8.1.
- Balance the single-phase loads on 3-phase distribution systems. the unbalanced load should be designed to not exceed 2% unbalance.
- In applications where there is no intent to adjust the speed of the motor, a full load, or off-control scheme, a soft starter should be provided.
- In applications where motor speeds are varied between 50% and 100% to adjust for system demands, VFDs typically having an efficiency of 95% to 98%, depending on the type of VFD provided (6- or 18-pulse, active frontend, low harmonic, etc.)

7.7 Emergency standby power to dedicated supplies

To meet Tier II requirements, a simple N+1 generator power plant architecture with one common MV switchboard and a basic electrical protection system (based on overcurrent and directional protection) should be considered.

Standby power systems designed to provide an alternate source of power if the normal source of power from the serving utility, should fail, shall be designed for reliability and in full compliance of NEC Article 701 or equivalent standards.

Additionally, the standby power supply system shall comply with the following Codes and Standards:

- IBC Chapter 27: Electrical
- NFPA 110: Emergency & Standby Power Systems
- NFPA 101: Life Safety Code
- NFPA 70: National Electrical Code

- NEC Article 445: Generators
- NEC Article 700, 701, & 702: Systems

The system should be capable of operating automatically or manually in an open transition or closed transition mode in conjunction with a load shedding system on the feeder breakers.

A separate custom control panel shall be provided to house the controls including a PLC, touchscreen and other control equipment. Hot Standby PLC equipment should be used, along with a control power transfer scheme to achieve high availability and a redundant system.

7.8 Uninterruptible Power Supply (UPS)

The product of choice for improving the reliability of electric power is the UPS. The UPS conditions utility power so that essentially perfect voltage and current is supplied to the protected equipment, called the critical load. The UPS should include batteries (or other energy storage devices) that keep power flowing to the critical load when the utility fails

UPS power shall be provided during the emergency generator engine starting period to be a buffer between the generators and sensitive load equipment power. The operational goals shall remain the same regardless of UPS topology: The supply of uninterrupted power to sensitive, critical loads.

7.9 Local Distribution/Protection

A distribution system should be provided to all items requiring a power supply. The distribution system should comprise main and sub-main distribution panels, trunking distribution systems, LV distribution cables, voltage regulators and power factor correction equipment.

The designer shall endeavour to develop flexible, sustainable electrical systems by maximizing resources. Specifically, the electrical systems design consumes resources both during construction and throughout the life of the building. Although a large part of the designs is driven by certain sections of the codes, these codes also contain some flexibility that allows designers to use fewer resources. ASHRAE Standard 90.1-2016 could be considered as the most sustainability-driving standards.

The low voltage electrical switchgear system consisting of passive and active components should be designed with the following in focus:

- The passive components include steel frames, cover plates, barriers, and horizontal and vertical bus structures.
- The active components are critical and include power circuit breakers or fused devices, as these components are responsible for protection from overcurrent.

The low voltage switchboards should have a common grouping of "sealed type" circuit breakers in a common enclosure. The breakers should be directly connected to the bus and may be group mounted or individually mounted in their own compartment within the entire enclosure.

Interfaces enabling connectivity to motor, breaker status, predictive information, control (on/off) shall be provided. Ethernet infrastructure communication system should be used with Modbus TCP/IP as access point for data and allow messages to be sent over the Internet. All low voltage electrical switchgear equipment should be connected to enterprise software packages (SCADA, EMS, BMS, etc.).

The power circuit breakers should be designed utilizing space-age materials packaged in compact fashion, with digital trip units. Digitization is essential to have enhanced equipment connectivity. This allows access to more detailed levels of energy management data, which enables better monitoring of energy consumption.

Facilities for electrical metering & monitoring shall be embedded within a circuit breaker.

Remote mimic panels shall be provided for circuit breaker control operations to minimize arc flash exposure of personnel.

A coordination analysis should be performed to provide power equipment with the required protection and minimize service interruption under overload or short circuit conditions. This analysis shall evaluate the electrical system's protective devices including relays, fuses, circuit breakers, and the equipment to which they are applied.

7.9.1 Distribution Panelboards

The designer shall use the terms distribution or power panels for panels that feed other panels or large loads, lighting panels for lighting loads and lighting controls, and receptacle or branch panels for panels that feed socket outlets or smaller loads.

7.10 Harmonics

Harmonic distortions generally transmitted by nonlinear loads, such as Switch-mode power supplies (SMPS), variable speed motors and drives, photocopiers, personal computers, laser printers, fax machines, battery chargers, and UPSs should be corrected through a harmonic current mitigation strategy.

Since harmonic mitigation vary in complexity and cost and can be deployed individually or in combination, the strategy that makes the most sense for the facility should be implemented based on the loads it supports, budget, and the nature of the harmonic-related problems the facility will experience.

7.11 Metering Facilities

All distribution boards serving loads in excess of 50KW should be provided with sub-metering.

7.12 Motor Control Center (MCC)

Each MCC shall provide a compact, modular grouping for motor control and electrical distribution components. A MCC should be implemented where a central control point is needed to remotely operate multiple loads. Distributed control systems (DCS) or programmable logic controllers (PLCs) should be used to provide this control and data acquisition functionality.

Intelligent Motor Control Center (IMCC) having Intelligent devices imbedded in the MCC shall provide network communications and functionality that is not available on standard devices, such as network configuration, diagnostics, process information, and advanced protection for each unit. Networks with high bandwidths should be used to facilitate the transfer of large amount of information available. The IMCC supplier should be able to integrate with multiple networks.

Motor control equipment shall be designed to meet the provisions of the National Electrical Code (NEC), and code sections applying to industrial control devices Article 430 on motors shall be fully applied. Standards established by the National Electrical Manufacturers Association (NEMA) shall assist the designer in the proper selection of control equipment.

7.13 Lighting

7.13.1 General

In many instances, energy codes require lighting engineers to design lighting systems that meet prescribed power allowances, use daylighting controls, control spaces through occupancy, and specify and perform functional testing in their lighting designs. ASHRAE and the International Energy Conservation Code (IECC) both require spaces surrounded by ceiling height partitions (walls) to have an individual manual control (switches/dimmers). The control must be within the space or remote located with an indicator that identifies the space/area it serves. Exemptions to this requirement are for areas that must be continuously illuminated for safety/security and corridors or stairways used for means of egress.

The following design criteria are to be considered as minimum requirements which may be exceeded by local practices. The design engineer shall liaise closely with the Architect, Interior Designer & Specialist Lighting Consultants with regards to the selection and location of decorative luminaries and effects, both within and outside the building. The specialist lighting designer shall influence some of the building's glazing properties, shade controls and other features to improve the benefits of daylight harvesting.

7.13.2 Control Systems

Lighting is typically one of the largest energy loads in any building, and historically, it has been the hardest to control.

The networked lighting control system leverages the power of digitization and granularity to completely control a building's lighting system from a centralized location. Networked lighting control systems provide centralized control to ensure that all spaces can be optimized around energy savings and visual performance.

A scalable architectural lighting control system should be designed for the control of architectural lighting for use in the various architectural environments, from simple meeting rooms, retail spaces to networked systems in large venues such as the Conference and Exhibition Centre. The hardware shall include modular dimming and processing panels, options for programmable control stations, and versatile interface devices.

In public sitting and transition areas such as an Atrium lobby, dimming controls should be used to make these spaces unique and inviting. In the ballroom and specialty restaurant spaces, dimming should be used for mood setting. In meeting rooms and conference rooms, dimming should provide the flexibility needed to accommodate a variety of presentation media.

Types of Lighting Controls

- Phase-Control
  - As voltage decreases, lamp power and lumens decrease
    - Forward, reverse, or center phase
- 0-10 V
  - A simple electronic lighting control signaling systems

    * the control signal is a DC voltage that varies between zero and ten volts.

- DALI
  - Digital Addressable Lighting Interface
    * Individual fixture control possible
- DMX
  - Individual fixture control possible
    * Typically used in theatrical environments
- Zigbee
  - Wireless, individual fixture control possible
    * Used for a variety of industrial controls

Based on standardized application profiles, energy-harvesting wireless devices from different vendors that can seamlessly communicate with each other should be used in combination with the wired solution. This approach of open connectivity and interoperability will offer a complete solution of integrated wireless LED controls.

Lighting systems shall have control steps between 30% and 70%, which could be accomplished with a number of variations such as switching alternating lamp, dimming ballast/driver, or stepped ballast/driver with even illumination in the space. Lights in corridors, electrical/ mechanical rooms, public lobbies, washrooms, stairways, and storage rooms shall be exempted from this requirement.

Since LEDs essentially put a 'chip' in every luminaire, digital addressability and local control flexibility become standard features. In retail spaces the lighting specialist shall integrate all control features associated with lighting via a centralized intelligent control panel or a distributed system. Specific inclusions in such systems should include occupancy detection, dimming, daylight linking and scene-setting. Scene Controls should be used to operate several device series and set multiple levels or actions via one command and typically used to:

- Create mood lighting and quickly transform in-mall ambience.
- Schedule-switch or dim several lighting areas at a day's start or end.
- Enable/disable occupancy sensors when moving from working to nonworking hours' control.

All public area lighting requirements should have lighting control platforms that can be activated via a combination of manual switches and/or automatic controls via time scheduling, light level or occupancy sensors.

The residential aparments should be fitted with a Room Controller that integrates automatic and manual control of lighting and enable auto-configuration to meet energy codes and provide intuitive, energy-saving lighting controls. Devices such as digital wall switches, vacancy/occupancy sensors, and daylight sensors should be plugged into the Room Controller and auto-configured to default operation. A Bluetooth module that allows dimming and daylight functions to be managed with the use of a smartphone app may be considered.

The lighting control technologies, including those that integrate occupancy recognition shall be tied into the property building management system or a micro-processor based centralized programmable lighting control. The designer shall endeavor to apply open protocol controls and avoid the limitations of proprietary control systems. In the spaces where occupancy recognition is applied, dual technology sensors and controls (Infrared and ultrasound) shall be applied to eliminate false occupancy sensing and ensuring proper systems operation and function. When an automatic time switch control device is used, an override should be located in a readily accessible space with a maximum override up to 2 hours.

Occupancy sensors are required in all conference/meeting rooms, employee lunch and break rooms, private offices, storage rooms, janitorial closets, and other spaces $28m^2$ or smaller enclosed by ceiling height partitions.

Lighting circuits should be monitored via a central location to ensure everything is functioning correctly and that circuits that fail can immediately be identified.

7.13.3 Daylight Harvesting

Designers could incorporate into building design to provide large open spaces and the generous use of glass facades for the sake of natural daylighting. Fire rated glass technology makes both natural daylighting and fire containment attainable. These glazing products could enhance the interior spaces and allow the more efficient use of resources by allowing daylight to extend deeper into the building.

Daylighting sources fall into two general categories: sidelighting and toplighting. For buildings with long, shallow floor dimensions, it would be feasible to provide daylight via sidelighting from windows or curtain walls. Toplighting which involves skylights could light more than 90% of the building area delivering light more deeply and evenly. In toplighting systems, direct sunlight must be difused to prevent glare

Daylighting shall be effectively integrated with the artificial interior lighting system. Daylighting control and automation systems shall sense the amount of natural daylighting and dim or shut off artificial lighting in the day-lit space, and provide the designed amount of work area illumination. The designer shall identify lighting dimming technologies that have built-in provisions for daylight control schemes. Guidance provided in the current version of IECC should be followed.

In spaces with plenty of natural light, daylight harvesting shall use continually monitoring and adjusting artificial and natural lighting to enhance occupants' comfort by eliminating variations in lighting levels.

In a Mixed-Use Building lobby with high ceilings Side-lighting may be considered to bringing daylight through vertical fenestrations. In this space, the window provides both view and daylighting.

A total light management system that includes automated shade control should form the lighting design strategy for addressing both energy and glare management. The best solution should combine automated shade control with solar-adaptive software, and cloudy-day/shadow sensors that allow the shading software to evaluate and respond to real-time daylight conditions. It is important to choose a shade manufacturer that offers a broad selection of fabrics, allowing the lighting designer/architect to choose the appropriate transmittance level based on the building's location and orientation.

Light level sensors should be used to determine lighting levels according to the sunlight filtering through windows in specific areas. Should a dimmer detect less than optimal artificial brightness in an area, it brightens lights to the desired brightness value. Similarly, when natural brightness begins to fade outdoors, the lighting control system should adjust the artificial light in small steps. Should they wish to, users can override these automatic light controls by means of a simple push-button.

Daylighting through top-lighting

The key to an effective daylighting system complete with skylights is balancing light and heat through skylight size, spacing and glazing selection.

Spectrally selective low e-glazing that allow high visible light transmission with substantially reduced heat transmission and UV and IR blocking capability skylights should be selected.

A software driven computer tool that helps building designers should be used to determine the optimum skylighting strategy that will achieve maximum lighting and HVAC energy savings for a Mixed-Use Building.

7.13.4 Exterior Lighting

Exterior and feature lighting at entrances, ornamental gardens and lobbies shall be provided through seamless integration and lighting transitions via a DMX interface gateway which should be compatible with different manufacturers' LED lights and theatrical equipment.

Building grounds lighting luminaires over 100 watts should have lamp efficacy of at least 60 lumens/Watt.

The exterior light fixtures should be designed to automatically switch-off based on daylight, and any decorative façade and landscape lighting must be automatically shut off between midnight and 6 am, with the exception of security lighting. The exterior lighting should also be controlled by a combination of photo sensor and astronomical time switch. The basic components of the exterior control system should seamlessly interface with the control strategy applied on the interior of the building for a total building lighting control solution.

The lighting control systems shall be tied to the building management system, with astronomical time switches, and photocells providing an input to the system. Dusk to dawn operation shall be accomplished with a photocell controlling all of the fixtures, or lighting fixtures that have integral photocells or a combination of the two approaches.

External lighting should be provided for all pedestrian and vehicular access routes to the building, service and car parking areas. Road lighting should be in accordance with local codes of practice.

Decorative lighting and/or flood lighting should be provided to the major elevations of the building, and to landscaped areas and signage. All external lighting should be time switch and photo-cell controlled, with manual override switches being provided in the Security Office. Separate controls should be provided for the following:

- Road, car parking and delivery area lighting
- Pedestrian area lighting and building entrances
- Decorative and flood lighting
- Illuminated signs and displays
- Water features/ artwork / sculpture
- Roof and external plant areas

Control of parking garage fixtures should have automatic shut-off in parking garages based on both occupancy and exposure to natural daylight. The occupant control should be by one or more devices that automatically reduce power of each fixture by a minimum of 30% when no activity is detected within a zone. The light fixtures shall be equipped with onboard occupant sensors. The onboard sensor should signal the fixture to reduce the light output to a preset level. The lighting designer should ensure to not jeopardize the safety of the garage occupants. Guidance notes described in the latest version of ASHRAE 90.1 shall be applied, so that the controls are triggered far enough in advance so that a car or pedestrian is not entering a dark area before the fixtures are triggered to react. Additional controls are required to automatically reduce lighting levels of fixtures located within

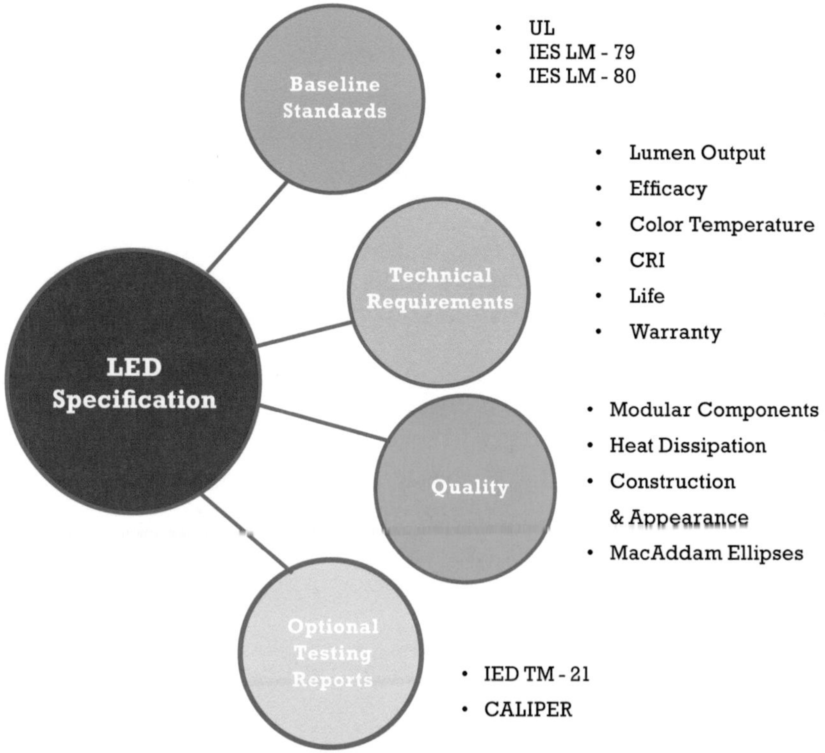

6m of a perimeter opening exposed to daylight. Similar to the indoor application of daylight control, parking garage daylight sensors should be installed to reduce the light output in response to daylight.

7.13.5 Luminaires

Taking into consideration the drive for energy efficiency, the design engineer shall advise the design team to maximize the use of energy saving long life luminaries. Solid-state lighting technology, such as Integrated LED lighting solutions that establish consistent color temperature in spaces, shall be implemented.

LEDs selected should have the capability for continuous dimming from 20% to 100%. However, linear fluorescents are required to have a minimum of four different illumination levels (20% to 40%, 50% to 70%, 80% to 85%, and 100%) for lamps greater than 13 W.

LED Recessed Troffers 3500 to 4000K lamps with electronic ballasts should be considered for office interiors. All digital ballasts shall comply with the Digital Addressable Lighting Interface (DALI) protocol.

In an environment such as retail store where color is important, the designer shall apply a high Color Rendering Index (CRI) to display the true colors of the displayed item.

LED luminaires shall be tested and certified to IMS LM 79 and LM-80 standards. LM-79 standard guidelines, and shall certify LED luminaires for light output, energy use, and color spectrum. LED luminaires shall have efficacies between 70 and 90 lumens/W. The minimum life of an LED luminaire shall be at least 70% lumen maintenance (i.e. L70) at 50,000 hours of operation, with a light loss factor of not more than 0.7.

7.13.6 Power over Ethernet (PoE) Lighting

The need for "smart" lighting in Mixed-Use buildings has set the stage for POE distributed networks. Since LEDs require low-voltage DC power in contrast with older lighting sources, which use AC power. Distribution of power and data on low-voltage cabling is possible using the same infrastructure deployed by the IT industry.

PoE IP networking safely allows electrical power and data to transmit concurrently on a single twisted-pair cable (Category 5e or above). Several standards bodies now provide for 15.4 W (PoE), 25.5 W (PoE+), 60 W (PoE+), and soon 100 W (PoE++). Many technologies today are using PoE to power devices.

LEDs are versatile, efficient, secure and able to support available wattages in POE networks. Each POE node will have a unique IP address, allowing the integration of LED fixtures. LED luminaires with bio-adaptive functionality are able to mimic natural daylight.

In a connected PoE system, each light will link to the facility's IT network. Lighting fixtures can become the focal point of an all IP network delivering intelligence, gathering and sharing data on occupancy, activity patterns, temperature, daylight levels, and other data.

PoE lighting deployment cabling should be based on the zone cabling standards structured cabling strategy where all system networks are converged within common pathways from the data rooms to consolidating points. The two deployment strategies for PoE lighting installations are:

- Centralized PoE switch zone cabling deployment
- Distributed PoE switch zone cabling deployment

Several requirements must be met to successfully converge a building automation and PoE lighting network. First, all building automation components must be IP-based or can be connected to an IP network through an adapter or converter. Second, the installed network cabling must be capable of handling PoE (generally 28 AWG – 22 AWG twisted pair copper cabling). Finally, to gain the full value of a converged building automation and PoE lighting network, a single pane of glass management software package would be required.

Technologies used in POE networks shall be in accordance with IEEE 802.3 standards, which specify the physical and data link layers for wired Ethernet, power sourcing equipment and devices.

7.14 Photovoltaics (PVs)

PV assemblies convert sunlight directly into electricity. PV systems encourage energy efficient design – such as daylighting strategies and building envelope improvements.

These systems are either ground mounted or integrated into the Mixed-Use Building roof or façade or take the form of an open structure. The main components are the PV panels or modules grouped together as an array.

7.15 Electric Energy Storage Systems

Energy storage is needed to store electricity, which is produced at times of low demand and low generation cost and from intermittent energy sources such as wind and solar power generation. It is released at times of high demand and high generation cost or when there is limited base generation capacity available.

Integration of power generated by decentralized renewable sources with the traditional power supply system is essential to mange overproduction of electricity during some hours of the day by generators based on renewable sources.

In order to capture this excess energy, an electric energy storage system must be incorporated within power systems. The energy storage can be additionally used for energy shifting either for peak looping or arbitrage as well as for providing ancillary services (e.g. power reserve). This multi-functionality can significantly improve the economic

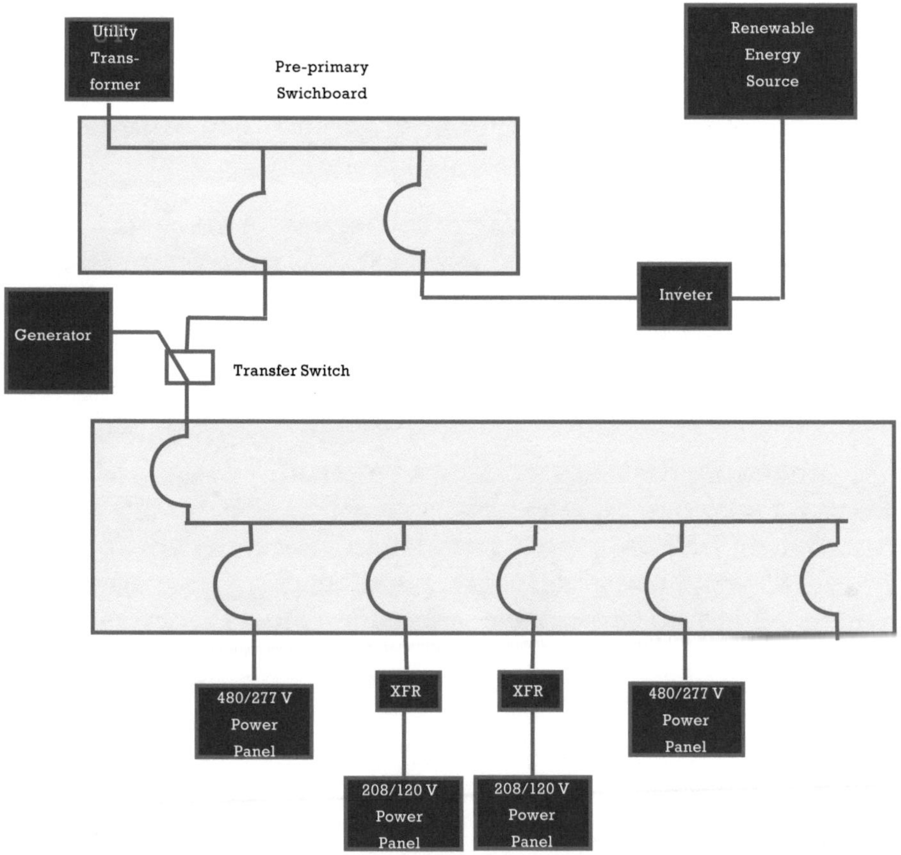

performance of these somewhat expensive technologies. Based on storage technology, a methodological framework to manage the energy surplus coming from renewables and CHP during low load and other operationally constrained conditions should be proposed.

7.15.1 Battery Energy Storage

Energy storage systems in combination with advanced power electronics (power electronics are often the interface between energy storage systems and the electrical grid) have a great technical role and lead to many financial benefits

The storage medium is the 'energy reservoir' that retains the potential energy within a storage device. Battery Energy Storage (BES) could be one of the considerations. A detailed analysis of energy storage benefits is essential. Following are some benefits:

- Cost Reduction or Revenue Increase of Bulk Energy Arbitrage
- Reduced Demand Charges
- Reduced Reliability-related Financial Losses
- Reduced Power Quality-related Financial Losses:
- Increased Revenue from Renewable Energy Sources

7.16 Integrating Alternative Power Generation Systems

Integrating alternative power systems such as PV and wind turbines, creates a great opportunity to reduce the carbon footprint, reducing dependency on fossil fuels and utility loads. While these systems are relatively straightforward, certain precautions and decisions must be made to ensure a safe and functional system by properly interfacing with the utility provider.

The two basic methods of connecting an alternative power system:

- Grid-tied
- Off-grid

The off-grid system relies 100% on power it generates, and this power is stored in battery cells. These systems are generally supplemented with a fuel-burning generator to ensure reliable power 24/7. The off-grid system hold merit for smaller buildings.

Grid-tied systems will provide all its generated power to the Mixed-Use Building with any additional power required being drawn from the electric power supply grid. Grid-tied systems can use batteries, but typically do not. In the case of a grid-tied system, if the load of the building is greater than the generated power of a PV of a wind turbine system, all the generated power will be used by the Mixed-Use Building. However, if the alternative power system would generate more electricity than the building load, the generated power should see the utility as load and backfeed power to it.

7.17 Vertical Transport & Escalators

Given the importance of reliable vertical transportation systems, newer approaches, based on data provided by the Building IoT network is collecting, analyzing, and acting on data enabling solutions to reduce energy consumption.

A Building IoT solution should provide facilities managers, quantitative data that helps them to determine whether their equipment is running as efficiently as possible.

For vertical transportation systems, a Building Internet of Things platform allows technicians to streamline and focus their maintenance and repair efforts. That translates to several compelling benefits for building owners and facility managers. Both system uptime and equipment life increase, due to more effective and targeted maintenance. This boosts building occupant satisfaction, which can translate to greater value for a property.

Section 8.0

Energy Management

8.1 Energy Conservation Measures

It has been emphasized in previous sections of this book the importance of designing and constructing buildings to the most current energy codes. The International Energy Conservation Code (IECC), developed by the International Code Council (ICC), and Standard 90.1, developed by the American Society of Heating, Refrigerating and Air Conditioning Engineers (ASHRAE) and the Illuminating Engineering Society (IES) are generally adopted to assist the energy efficient design and construction of new buildings.

8.1.1 How to Achieve Building Energy Goals

A Mixed-Use Building is a very dynamic operation. People come and go, The Sun shines on some windows in the morning and other windows in the afternoon, various meeting rooms are used at different times of the day, and physical occupancy of the guest rooms varies with the needs and whims of the individuals.

The goal is to improve the overall energy efficiency of any structure and reduce the energy needed to maintain a healthy, comfortable, and fully functioning indoor environment. Energy codes generally apply to:

- Components associated with the building thermal envelope
- Heating, ventilating, and cooling systems and equipment
- Lighting systems and equipment
- Water-heating systems and equipment

To effectively manage the initial cost for a project and be most effective, an integrated design process is essential. The project architect must ideally first employ sensible passive design strategies, then collaborate extensively with the MEP and Technology solutions designers, and other specialist consultants who will affect the performance of the building both in the design phase and after construction, to optimize the building design

8.1.2 Energy Management

Designing for energy efficiency can impact the look, feel, and function of the Mixed-Use building. For example, lighting and window design can impact cooling loads, window impact lighting etc.

Building Energy Management would require software, hardware and other services to implement intelligent monitoring, management, and control of energy and ensure the building's efficiency of operations. Energy efficiency, resource scarcity, and facility management efficiency will continue to be drivers for intelligent, responsive, networked, and adaptable energy management systems.

New advances in wireless technology has emerged as a viable and cost-effective method of achieving building control. The optimum solution for a given building will often be a combination of major wired and wireless technologies and standards that could be scaled to the most advanced building management system (BMS). In practice, these approaches often co-exist within a single facility. That's because each has its own advantages and disadvantages and using all the choices enables building owners to optimize a system to their particular needs and budgets.

Wired communication systems can coexist with wireless, and there may be a case where wired systems are still needed or preferred. Some reasons for using wired communication systems are:

- In new construction, running wires is not a significant extra expense.
- Some structures, because of their nature or materials used, may need wired systems to avoid problems with signal clarity.
- Very large facilities such as resorts may use both wireless and wired systems, running wiring between buildings and widely separated zones.
- For some building owners and contractors, wired systems are simply a preference based on training and experience.

Wireless technologies comply with green building regulations and guidelines such as LEED and may qualify for utility and government rebates. Not only does wireless enable systems that use less energy, but it also assists in reducing the use of expensive resources, particularly copper which is used in wiring.

Below are some critical issues that need to be addressed by vendors to help decision-makers find and implement a building automation solution that is right for them:

- Is it an open solution?
- Wireless vs. wired
- Simple. Ease of installation and use.
- Can it manage more than lighting?
- Smart. Does it provide useful reporting?
- Scalable. Can it grow with you?
- Code compliance
- Does it integrate with Open ADR (Automated Demand Response)?

Energy data collection is an important aspect in achieving energy management goals. The Internet of Things (IoT) provides a platform for centralized monitoring and control. Data visualization through dashboards with basic trending and alarm capabilities should be provided. These dashboards should uncover a complete energy picture, not just a snapshot.

All of the energy consuming assets should be networked and operate in an open framework where dashboards provide free and ubiquitous connectivity. Energy dashboards should display in real time how to tune or change the use of energy to deliver just what the facility needs and realize significant energy savings. The energy dashboard should be designed to provide easy-to-understand graphics based on granular collection of energy use data at the individual circuit level.

8.1.3 Energy Audits

Overall energy losses in a Mixed-Use building can result from losses due to designs that do not incorporate energy efficient specifications; operations that run on inefficient methods; and poor or non-energy efficiency-conscious maintenance program. Reducing these losses will substantially increase the building's efficiency, but data is required to identify and quantify the losses and subsequently suggest suitable techno-economic solutions to minimize the losses. This data can be acquired through Energy Audits.

Energy audit is designed to determine where, when, why and how energy is being used in existing buildings. This information could then be used to identify opportunities to improve efficiency, decrease energy costs and reduce greenhouse gas emissions. Energy audits are used to verify the effectiveness of energy management opportunities (EMOs) after they have been implemented.;

a) **Types of energy audits (walk-through and detailed energy audits)**

Depending on the level of detail on the collected information, energy audits might be distinguished into two types, the walkthrough and the extended audits. Walk-through energy audits assess building energy consumption and relevant costs on the basis of energy bills invoices and a short on-site autopsy.

Detailed - diagnostic energy audits require a more detailed recording and analysis of energy and other data. The energy consumption is disaggregated in different end-uses (e.g. cooling, heating, lighting, etc.) and the different factors that affect that end-use are presented and analysed (e.g. occupancy, climatic conditions, food and beverage business, etc.). All the cost and benefits for the energy saving opportunities that meet the criteria and requirements of the end-energy Mixed-Use Building administration are determined. A list for potential capital-intensive energy investments requiring more detailed data acquisition and processing is also provided together with an estimation of the associated costs and benefits.

The energy use intensity of the building compares current energy performance with internal historical performance and external established benchmarks. As shown below:

BENCHMARKS FOR UTILITY CONSUMPTION: Large Residential Apartmentss (more than 150 apartments) with full air conditioning, laundry and indoor pool without gardens	
Electricity	**<165 KWh/SQM/Year**
Gas, Fuel and Steam	**<100 KWh/SQM/Year**
Total	**<265 KWh/SQM/Year**
Water	**<600 Liters per guest**

BENCHMARKS FOR UTILITY CONSUMPTION: Medium Residential Apartmentss (50 to 150 apartments) with full air conditioning, but without laundry, indoor pool or gardens	
Electricity	**<70 KWh/SQM/Year**
Gas, Fuel and Steam	**<80 KWh/SQM/Year**
Total	**<150 KWh/SQM/Year**
Water	**<440 Liters per guest**

BENCHMARKS FOR BEDROOMS	
Electricity	**<15 KWh/OCRM/day**
Water	**300 Liters/guest/day**

BENCHMARKS FOR KITCHENS	
Electricity for Cooking	**<3 KWh/Cover**
Electricity for Lighting, ventilation and cold-rooms	**<1 KWh/Cover**
Water (hot and cold)	**27 Liters/Cover**

BENCHMARKS FOR LAUNDRY	
Electricity	**<0.65 KWh/Kg**
Steam	**<1.6 KWh/Kg**
Water (hot and cold) - without recovery	**<20 Liter/Kg**
Water (hot and cold) - with recovery	**<13 Liter/Kg**

Note:

Since Mixed-Use Buildings differ in many ways, when comparing utility consumption figures with the benchmarks, correction factors must be applied to take account of weather conditions and occupancy levels, as well as the presence of facilities such as laundry, indoor pool and health club, each of which has a significant impact on energy and water use.

8.1.4 Energy Metering

With advances in energy metering and information systems resulting in increased functionality at lower costs, obtaining these data in a cost-effective manner is now a standard practice. The application of meters to Mixed-Use buildings and energy-intensive equipment provides facility management with real-time information on how much energy has been or is being used.

Energy metering facilities should be provided to allow the energy and thermal performance of the building to be recorded and to establish baselines for performance and management, energy savings verification and Heating/Cooling plant optimization and control.

Cellular wireless wide-area network (WWAN) technology shall be applied where it is identified to provide distinct business and technical advantages over conventional radio communication networks. By enabling two-way communications between meters and utilities provider and providing accurate information about the status of the utility supply system, smart meters play an important role in energy conservation.

"Smart" meters shall use Advanced Metering Infrastructure (AMI) to allow for two-way remote communications with meters, using either point-to-point cellular WWAN technology deployed at the meter, or "fixed-network" solutions with cellular gateways connecting groups of meters using low-power RF mesh networks.

Smart metering shall typically include meter data management (MDM) software and services that provide utilities with end-to-end capabilities to remotely read meters and integrate data with back-end systems.

Wireless communication networks shall be configured for network security issues related to data integrity and information security when connected to standard data networks. To achieve these network security requirements, the communication protocols should include appropriate levels of information assurance and security, through control protocols that have been extended with a set of network layer security messages that provide data confidentiality and integrity, device authentication, data hiding, and user authentication.

Central energy plants that meter both chilled-water and heated-water circulation systems, energy shall be determined as the product of the fluid's mass flow and the corresponding temperature differential through the system being metered. Positive displacement, differential pressure and velocity meters may be used in heated and chilled water circulation systems.

a) Sub-Metering

The sub-metering of the Mixed-Use building provides the operations and maintenance transparency necessary to enable more efficient management of energy resources. In addition, sub-metering can drive behavioural change related to energy conservation and advance real-time building interaction with a Smart Grid. Each of these potential benefits can dramatically improve building performance and lead to reduced resource consumption.

While sub-meters by themselves have no direct impact on resource use, the data they capture informs real-time energy performance, can pinpoint performance variations over time or relative to feeds into building automation systems that drive continuous operational improvements, and provides the information needed to encourage behavioural and operational changes by facility managers and staff.

Sub-metering systems should be carefully designed to meet stated operational criteria and objectives, i.e., data analysis and operations management requirements must guide sub-metering hardware and software selections and specifications of the system configuration.

Sub-metering designs shall include the specification of the level of sub-metering required, the types of data to be collected and used to reduce consumption, and the software, analysis, and communication tools necessary to manage energy usage by systems, in order to provide information to facility management and/or to control the building's energy use directly.

Sub meters shall interface with the building automation system, not only making it possible to integrate the control logic with the meter, but also providing access to the building automation system tools for trend analysis, report generation, and user information display.

I. All meters shall be clearly labelled and easily accessible.

II. Separate meters shall be provided for the following major building systems

- Heating Ventilation & Air Conditioning
- Guestrooms
- Kitchens
- Laundry
- Fitness Centre & Swimming Pools
- Elevators & Escalators

b) Meter/Sub-Meter Performance Metrics and Attributes

Performance measures for meters and sub-meters shall include but not limited to the metrics listed below:

- Accuracy
- Precision/repeatability
- Turndown ratio
- Ease of installation
- Ongoing operations and maintenance
- Installation versus capital cost

c) Communication Networks and Data Storage Requirements

Regardless of the meter type, once sub-metering data is collected it must be transmitted via a communication network to be processed, stored, and used. The type of sub-metering plan with its requisite data requirements should define the communication network appropriate for the application.

The options for automated metering communications include phone modem, local area network (LAN), building automation system, radio frequency (RF), and wireless network.

Metered energy data should be analysed at regular intervals (e.g., every 15 minutes is standard in the U.S.) and may incorporate other relevant information:

- Meter location
- Usage points (e.g., lighting, HVAC)
- Kilowatt-hours of consumption
- Highest kilowatt demand
- Voltage
- Real and reactive power
- Hourly weather information (temperature, humidity)
- Local currency values of consumption, peak charge, access charge, taxes, and delivery charges
- Calculated values for load factor, per square foot or per square meter use
- Heating or cooling degree days

The above-mentioned information shall be integrated with software analysis capabilities. The building's advanced meter, sub-meters, and building automation systems should all provide data to an Energy Information System (EIS) which should have analysis tools ranging from spreadsheets on local computers to integrated systems with analysis capabilities, trend identification, report generation, and graphical displays.

Dashboards with energy use applications, equipped with features and interfaces that are tailored to the needs of the Mixed-Use Building operator shall be provided. They shall include real-time load profiles and baseline comparison profiles that correct for weather and other external conditions, so that the facility operator can accurately determine the impact of energy conservation initiatives.

8.2 Building Management System (BMS)

Building control systems play an important part in the operation of a building and determine whether many of the green design aspects included in the original plan function as intended. Controls for HVAC and related systems have evolved over the years, but in general, they can be described as either distributed (local) or centralized. Local controls are generally packaged devices that are provided with the equipment. A BMS, on the other hand, is a form of central control capable of coordinating local control operation and controlling HVAC and other systems (e.g., life-safety, lighting, water distribution, and security from a central location. BMS architecture should be capable of encompassing HVAC and other systems integration through multiple protocols into an integrated system.

In the new data-driven lifecycle of modern Mixed-Use buildings, BMS and controls services is an effective bridge between design, construction, and operation. Optimal operating sequences and relationships shall be built into the "smarts" of the building, and performance requirements shall be made software enforceable.

To be able to adapt to the needs of the facility operator and occupants over time without costly hardware changes software overhauls, the contracts for hardware installation shall be separated from the contracts for software that include but not limited to controls, analytics, and workflow management.

The building should be equipped with a BMS, providing comprehensive building automation functionality

- To maintain desired thermal comfort and lighting levels for guests
- To monitor thermal and electrical energy consumption
- To monitor and control all major HVAC plant
- To monitor facility security and maintenance issues
- To operate the HVAC systems on programmed schedules and use weather conditions to predictively maintain operating conditions
- To provide trending and long-term monitoring of facility use for benchmarking and future expansion
- CO_2 level monitoring for all HVAC Zones and control of fresh air supply to respective zones
- HVAC systems control under "fire alarm" mode.
- Control, monitoring and audible alarm functions associated with all Mechanical Services.
- Monitoring of space temperatures, relative humidity in all zones and areas served by central air handling units or individual fan coil unit systems.
- Monitoring of the alarms, alarm printing and acknowledgement

a) Control sequences for HVAC systems

- The sequence of operation is one of the most important design aspects of any HVAC system. Design a proper sequence using the following steps:
- Create a flow diagram of the system
- Categorize the purpose of the equipment
- Identify the required inputs and outputs
- List any code required functions of the system
- Confirm the owner's operational requirements and expectations
- Develop a list of points
- Identify the setpoints
- Work through the actions and functional responses
- Identify failure scenarios
- Review the sequence

8.3 Thermal chilled-water storage

Many modern high-rise towers include a multilevel below grade parking structure. Based on the requirement for a fire-water storage tank on the lowest level, there is often an opportunity to increase the size of the tank and provide a thermal separation by adding a chilled-water thermal vault. The size of the water vault must be coordinated with the structural mat foundation to assure proper sizing and location.

8.4 Renewable Energy

To meet a minimum level of overall energy performance in a building, the design team should use a holistic design approach that minimize heating and cooling demands by integrating advanced high-efficiency building systems that utilize ***renewable energy*** sources, and make the best use of any fossil fuels through application of highly efficient energy conversion technologies. Sophisticated design methods that include the technical and economic modelling of energy systems – demand and supply is of utmost importance.

The design team should conduct an analysis that is focused on adding onsite renewable energy, including wind and solar.

Renewable energy technologies collect renewable energy and transform it into a form that will, ultimately, meet energy service demands. Some of the leading renewable technologies and typical city resources are summarized

a) Building-integrated photovoltaic (BIPV) panels may be connected to the façade to provide shading and/or equipment screening while generating electricity. Current PV technologies, such as glazing and roofing systems with integrated PV collectors, are other possibilities for generating electricity without having an impact on the structure's architectural aesthetics.

b) The most efficient way of using wind energy is to use a large turbine located away from the building. These turbines often produce more energy than is required by the building and therefore provide the opportunity of exporting electricity to the grid.

c) Consider the use of heat pumps – generally for every kilowatt of energy fed into a heat pump 3kW of energy are produced. If the two units are free (i.e. from groundwater, river water, air or other sources), then high efficiencies are obtained.

Ground source heat pumps (GSHPs), coupled with certain types of low-temperature distribution system would be an efficient and environmentally friendly heating and cooling technology for many climates.

d) Co-generation of heat and electricity that converts fuel into heat, which is then used to produce steam and electricity, and the waste heat used for heating demands is a possibility.

The best mix of renewable energy technologies is that which provides the required amounts of energy at least cost, while meeting other constraints such as air pollution emission and air quality.

8.5 Miscellaneous Energy Saving Devices

a) Variable Frequency Drive (VFD)

A VFD is specified to reduce operational cost for pumps, fans, compressors, or any similar equipment with variable load profiles that may be found in a typical building. To effectively specify a VFD, it is essential to go back to basics and methodically work through a few key steps:

- Understand the load (operating power, torque, and speed characteristics)
- Understand duty cycle (what percentage of operation at 100% load, 50% load, etc.)

Once information from the above two items have been established, determine what outcome one is trying to accomplish by using a VFD (energy savings, soft start, controllability, etc.)

Excluding constant horsepower and constant speed/torque loads, typical loads that can take advantage of VFDs can generally be divided into two categories:

- Variable speed, variable torque (fans, blowers, and centrifugal pumps)
- Variable speed, constant torque (positive displacement loads such as screw compressors, reciprocating compressors, or elevators).

b) AC Induction Motor Efficiency

Substantial reductions in energy and operational costs can be achieved through the use of energy-efficient electric motors. Economic and operational factors should be considered when motor purchase decisions are being made.

Energy-efficient motors are truly premium motors. The efficiency gains are obtained through the use of refined design, better materials, and improved construction. The price premium for an energy-efficient motor could be 15 to 30 percent above the cost of a standard motor. Durable and reliable energy-efficient motors could be extremely cost effective with simple paybacks on investment of less than 2 years.

It is critical that motor efficiency comparisons be made using a uniform product testing methodology. There is no single standard efficiency testing method that is used throughout the industry. The most common standards are::

- IEEE 112 - 2017
- IEC 34-2

Other things being equal, seek to maximize efficiency while minimizing the purchase price.

Energy-efficient motors are a worthwhile investment in all size, speed, and enclosure classifications. In general, higher speed motors and motors with open enclosures tend to have slightly higher efficiencies than low-speed or totally-enclosed fan-cooled units. In all cases, however, the energy-efficient motors offer significant efficiency improvements, and hence energy and cost savings, when compared with the standard-efficiency models.

Energy-efficient motors typically operate cooler than their standard efficiency counterpart? Lower operating temperatures translate into increased motor, insulation, and bearing life.

c) Permanent Magnetic Alternating Current (PMAC) motors

The functional benefits of PMAC motors have led premium cooling tower models to use these motors. The manufacturer takes advantage of the low-speed performance characteristics of PMAC motors and eliminated the use of fan belt drive/reduction gearboxes typical in such designs.

This type of application, where the operational speed of the load is dramatically less than the standard base speed of an induction motor (1800 rpm, 3600 rpm, etc.) and belt drives or mechanical gearboxes end up being required, exemplifies where this type of motor reduces operating costs.

Section 9.0

Water Management

9.1 General

Water is one of the basic sources of all life on earth. It is fundamental to food production, public health, and the health of all living species. Fresh water is one of the most fragile of the world's resources.

As the world's population continues to grow and cities and suburbs expand, with more buildings to accommodate the growth, a major challenge to manage and protect valuable water resources becomes critical. This is becoming especially critical under the added pressure of climate change as certain locations are experiencing moderate drought or abnormally dry conditions.

These conditions place enormous pressure on potable-water storage and water supply networks. This not only creates water-supply uncertainties, but also brings about a realization in the broader community that water is an extremely valuable and finite resource that should not be taken for granted.

As the availability of clean freshwater resources is diminished, it is essential that wastewater and storm water should be viewed as alternative and valuable sources of water. Consequently, rainwater collection and reuse and the reuse of highly treated wastewater effluent are practical solutions. They are most-commonly provided for non-potable end uses; potable reuse is also gaining traction.

9.2 Water Consumption

The commercial and institutional sector is the second largest consumer of publicly supplied water in the U.S., accounting for 17 percent of the withdrawals from public water supplies. This sector includes a variety of facility types such as Mixed-Use Buildings, restaurants, office buildings, schools, hospitals, laboratories, and government and military institutions. Each facility type has different water use patterns depending on its function.

Each building type faces unique challenges and has specific areas where the greatest reductions can be made, but significant water savings can be achieved indoors and out through improvements in equipment and operational practices.

For example, water used in office buildings accounts for approximately 9% of the total water use in commercial and institutional facilities in the United States. The three largest uses of water in office buildings are restrooms, heating and cooling, and landscaping.

9.2.1 Assessing Facility Water Use

Understanding how water is used within a facility is critical for the water management planning process. A water assessment provides a comprehensive account of all known water uses at the facility. Assessing facility water use incorporates the following steps:

- Gathering readily available information
- Establishing a water use baseline
- Inventorying major water-using fixtures, equipment, systems, and processes

Water sources can be classified as follows:

Potable water: Water that is of sufficient quality for human consumption and that is obtained from public water systems or from natural freshwater sources, such as lakes, streams, and aquifers that are classified, permitted, and approved for human consumption.

Non-potable water: Water that is obtained from natural freshwater sources that is not of sufficient quality for human consumption and has not been properly treated, permitted, or approved for human consumption.

Onsite alternative water: Water that is not obtained from a surface water source, groundwater source, nor purchased reclaimed water from a third party. It can include rainwater or storm water harvested on site, sump pump water harvesting, gray water, air-cooling condensate, reject water from water purification systems, water reclaimed on site, or water derived from other water reuse strategies.

Purchased reclaimed water: Wastewater treatment plant effluent purchased from a third party that has been diverted for beneficial uses, such as irrigation, that substitute the use of an existing freshwater source.

9.3 Water Conservation

Water is scarce resource besides being expensive. Use of alternate water sources is one way to meet the demand. Reduce, reuse, reclaim and recycle is the future of sustainable urban water management systems. Effectiveness in water saving, equity in water sharing and delivery efficiency are essential for the sustainable use of available water resources.

Dish and clothes washing machines significantly contribute to the amount of water use in a residential apartment complex. Energy Star clothing washing machines can use up to 50% less water than a typical washing machine, as well as contribute to potential energy savings. Energy Star dish washing machines can contribute up to a 25% savings in water. These two technologies offer significant water savings..

9.3.1 Water Balance Modelling

To establish where water savings can be made, a detailed water-balance model of the project should be carried out during the design phase. This process will identify all possible water savings by incorporating uncomplicated water-efficiency strategies into the development design. Using this approach, the use of water-efficient fixtures and water-conservation design strategies can reduce an average retail or commercial development's overall water use.

Analysis of the water-balance model at the outset of the project will reveal any opportunities for alternate water supply, water conservation, and water recycling. Water-balance modelling will also reveal the availability of rainwater and wastewater that could be captured and treated for reuse.

All aspects of facility operations that involve water should be investigated and potential sources of water identified. Both internal and external water sources should be considered. A water balance could be used to quantify both the reuse demands and the potential reuse supplies, as well as the timing of those supplies, to combine the two in the most economical and practical way.

An overall strategy should then be developed based on matching quantities of demands and potential reuse supplies. The strategy should incorporates an integrated approach for deploying water-use reduction along with various reuse schemes.

Mixed – Use building designers should consider the water conservation options discussed in this publication and implement those which are cost effective. Since production, treatment, and transportation of water require energy and chemical usage, building designer's efforts to conserve water also help to support the installation's pollution prevention goals.

Besides the benefit of securing the world's water supply for the future, other important benefits can be derived from water conservation. Proper water management can lead to substantial financial savings. When water is conserved, energy savings are often observed due to lessened energy demands for treating, heating, cooling, and transporting the water. Pollution prevention benefits are realized in two ways:

- Reduced energy means reduced air pollution
- Less water treatment means less chemical usage

To be water-efficient, the ***Reduce, Replace*** **and** ***Reuse***(3Rs) approach should be adopted in the initial stage of planning and designing of a building. Effectively implementing the 3Rs can result in substantial savings in water, as well as energy, therefore reducing the environmental impact of both water discharge and the need to pump water over long distances.

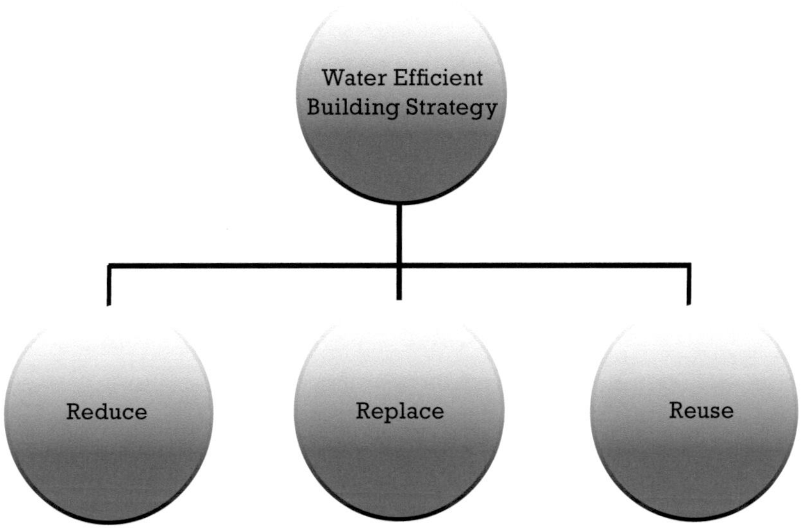

Innovative new technologies and timely new product development in water efficient fixtures has delivered digital hardware and software that enable designers to comply with key performance objectives of water efficiency.

a) Water Reuse

Potable-quality drinking water must be supplied for human consumption and food preparation, water used for washing hands, showers, and where close and intimate contact with water is required.

Water reuse from or to other sources is another alternative. For instance, site-collected storm water delivered to the landscape garden by sub-surface drip irrigation may only require simple filtration to remove particles that may block the irrigation drippers. On the other hand, harvested storm water runoff from the forecourt area and rainwater collected from the main building roof is passively treated through media filtration at the source, eliminating the need for a mechanical plant, reducing the overall energy demand, and providing water suitable for toilet flushing. Whereas, recycled water for use in cooling towers may require a much higher level of treatment including microfiltration and disinfection to remove pathogens.

Water stream	Definition
Potable water	Also referred to as drinking water or domestic water. A stream of high-quality water, which is provided at a cost per unit volume of water used by the utility provider.
Rainwater	All rainfall runoff from a given catchment. The combination of both roofwater and stormwater.
Roofwater	Water collected from the hard surfaces of roofs. Roofwater is relatively straightforward to collect and is suitable for nonpotable use with little pretreatment
Stormwater	Water collected from ground-level hardstands. It is generally more contaminated than roofwater, not only because of its contact with the surfaces of roads and parking lots or decks (e.g., wear and tear of tires, oil, and grease) but also because stormwater drains to the lowest point, which is where dust and litter settles.
Gray-water	Usually a high-volume, relatively low-contamination wastewater stream from showers and sinks and laundries (does not include wastewater from toilets and urinals).
Blackwater	This represents the most highly contaminated wastewater stream from toilets, urinals, and kitchens. Blackwater can also mean the entire wastewater stream leaving a development (blackwater and gray-water combined).

Definitions of water streams

There are also environmental advantages to consider. The use of reclaimed water ensures the transfer of nutrients to beneficial uses rather than discharging them into receiving waterways.

Reuse may require the installation of dual pipework either to separate wastewater streams, such as where gray-water alone is being treated and recycled, or to provide separate supply lines so reuse water can be used for other purposes, such as toilet flushing. This duplication of infrastructure and provision of treatment systems can be costly, and therefore the benefits of water-reuse schemes should be carefully evaluated. They are generally more likely to be implemented where cost is not the only factor, such as where the true cost of water is only realized when it is not available.

The two major plumbing codes in use in the United States are Uniform Plumbing Code (UPC) and International Plumbing Code (IPC). These codes cover aspects of alternate water sources and non-potable applications including gray-water reuse and rainwater harvesting systems.

b) Reduce Consumption

Under the Reduce approach, the design team should plan and install a monitoring system so that water consumption can be tracked and reviewed, adopt low pressure water system and select water efficient systems for cooling, irrigation, etc, employ high water efficient labelled products and design a leak-free system. Besides reducing water usage, designers should plan and design the building to Replace the use of potable water by substituting potable water with other sources of water to supply the water demands. The aim should be to provide water sources that are less reliant on weather and rainfall events and, therefore, more likely to be available during times of water shortage.

In mixed-use buildings restrooms alone can potentially account for 25% of a given building's water consumption, there is a big opportunity to track and conserve this amount. A low-consumption or high efficiency plumbing fixture is the most popular and obvious place to begin. These include water-efficient dual-flush toilets, low-flow showerheads, low-flush urinals, waterless urinals, low-flow faucets, and flow restrictors.

Technology

New technology helps better control and monitor the flow of water through the water delivery systems. This technology would identify potential leaks or isolate those pumps and valves that are starting to gradually fail before they stop working completely.

Water savings can be obtained through water pressure reduction. Automation systems can bring a large amount of savings when applying variable frequency drives in water pumps. automation systems automatically adjust the speed of VFDs to find the optimum speed at which to run them at all times.

Pressure reduction can be achieved by installing intermediate tanks at suitable levels in high-rise buildings. These intermediate tanks installed at levels lower than the high-level water tank will serve water fittings at designated floors in the tall building.

c) Replace

Besides reducing water usage, designers should plan and design the building facility to Replace the use of potable water by substituting potable water rainwater wherever feasible for their non-potable usage such as irrigation, general washing, cooling tower, etc.

Other uses in commercial buildings that are further removed from human contact may be supplied non-potable water such as irrigation, toilet flushing, cooling towers, boiler supply, and others. In each case, an assessment of the risks in terms of the chemical quality of the water and the likely presence of pathogens should be performed to ensure that appropriate treatment is performed prior to reuse. Recycled water is becoming more prevalent with many municipalities offering recycled water supplies.

The costliest, but most highly effective way to reduce water usage at mixed-use building facilities is to implement a wastewater recycling system. The type of wastewater treatment system will depend on the configuration of the building facility. The amount of treatment required will depend on the contamination level of the water source, the quality requirements of the water demand, and the likelihood of direct human contact.

9.4 Rain Water Harvesting

Building siting, geographic location, and weather conditions (i.e., average rainfall rate) could have a large impact on the feasibility of using a rainwater harvesting system. Rather than getting rid of storm water as quickly as possible, a sustainable approach to storm water management should include ways to harvest it on-site and using it for irrigation and groundwater recharge.

Irrigation demands are most-often highest during the hotter months. in climates with a higher summer rainfall, collecting rainwater to provide irrigation over summer is more efficient; the tank is likely to be drawn down through irrigation between rainfall events, thereby maximizing capture, resulting in a much smaller tank size. This also applies to rainwater collection for cooling tower water supply demands, which are also much higher during the hotter months.

9.5 Landscape Irrigation

Water for landscaping use is often the most pervasive consumer of water on any type of building and up to as much as 70% for some building types.

To maximize irrigation efficiencies, landscape designers should group plants with similar water needs together and use native or non-native vegetation appropriate for site conditions and climate. Plumbing design engineers should specify efficient irrigation systems and choose climate-based and "smart" controllers for irrigation systems that respond to environmental conditions and adjust appropriately.

The newest generation of smart controllers for irrigation systems can connect to the internet and interface with smart phones, so system operators can access them anytime from anywhere. The controllers should allow facility management staff to operate systems by setting specific programs or having pre-set programs run based on soil and weather conditions. For instance, controllers can account for such weather elements as wind, rain, temperature, and humidity, which keeps turf and other vegetation healthy and minimizes water use.

Smart controllers should have pre-set programs based on soil profiles, which allows the systems to monitor the rate at which water leaches and absorbs. As a result, the system will only water the grounds when the soil needs hydration.

Smart controllers also help facilities stay within watering guidelines in regions in which water use is restricted based on the time of day or an amount per week. Water-conservation initiatives have become so strict that some states now use water police and water courts to make sure residents and businesses comply with the regulations.

Wireless weather and wireless soil moisture sensors provide real time weather and moisture content data that allows for immediate adjustments. The smart controller using a local weather sensor should automatically download weather data from the internet.

To accurately connect moisture sensors into irrigation systems, managers and their teams must understand the way the technology works and the most appropriate location. If workers install it too high or too low, it will be ineffective.

Implementing flow sensors into a facility's irrigation system is one of the most beneficial ways of optimizing its efficiency. Flow sensors help system operators determine the amount of water a system uses regularly, as well as the rate of pumping.

These sensors control the amount of water used, and they can assist operators identify potential problems in the system. For example, the sensors will signal the controller to automatically shut down the flow if an overflow or leak occurs.

Flow sensors also assist in gathering essential data for facility managers seeking to make more informed decisions about their landscapes' irrigation needs. For instance, operators using systems with flow sensors can determine the amount of water the facility expects to use, which can help managers plan for irrigation costs for the following year.

Pressure regulation on spray heads prevent runoff and put down only the amount of water needed and saves water by controlling how much water is released. They fit inside a sprinkler to control the water pressure.

9.6 Water Tracking

With so many functions requiring water throughout a typical commercial building, it is essential that a localized and robust method of tracking water consumption needs to be implemented. This is where sub-metering and enhanced data-retrieval technologies are required. Sub-metering would allow the facility operations staff to pinpoint issues more easily. In order to analyse the separate building functions, sub-metering and enhanced data retrieval would more precisely track water consumption.

Providing water sub-metering on separate tenant spaces gives the facility manager the ability to individually track water consumption from tenant to tenant or space to space. Installing sub-meters on HVAC equipment gives the facility manager the ability to individually track this water-intensive equipment, and to ensure the equipment and systems are functioning as designed without any issues such as leaks or blockages.

Landscape irrigation is an area where water can be more effectively tracked.

Water intensive areas such as Kitchens, laundry, or any large water-using process must be sub-metered. Water reuse systems such as gray-water, storm water, and HVAC condensate collection systems should be sub-metered. Metering these systems will allow the facility manger to not only track what they are using, but also track what they are gaining or offsetting.

Smart meters are taking center stage as solutions that can increase water efficiency. Flow meters for commercial buildings should have many features in its electronics. Either electromagnetic meters or Ultrasonic water meters should be considered for the different applications. Meters with cutting-edge technologies that operate over large ranges are needed for more efficient water use. It is essential that an operator is able to read the meter on a cell phone with the use of a mobile application with Bluetooth technology.

9.6.1 Data retrieval and evaluation

Remote metering or automatic meter reading (AMR) capabilities must allow all metered information to be easily collected and analyzed. Remote metering and AMR allows for all of the information to be compiled in a single location and ensures that design conditions are being met.

Meter data-collection systems should be available to track daily and even hourly usage of water at meters and sub-meters. Data storage systems should have the capability of storing data for prolonged periods. This is essential for building maintenance, systems commissioning, measurement and verification, and comparing certain functions from time to time to make sure operation is consistent.

Smart sensors that can connect to the Internet and then share data about how much water is being pumped through the plumbing distribution systems, how much pressure they are using at different sections, and how active valves, pumps, and filters are, can make a big difference in how the facility deliver water.

These sensors can monitor possible leaks in a system, sending alerts to facility management when the water being pumped through a section of the pipe system differs from the amount of water reaching the end-users. IoT sensors can also help facility management better monitor the quality of the water being delivered, more accurately measure how much water their end-users are consuming, and monitor the health of their infrastructure.

Dashboards are essential for commercial building tracking. They would be monitors located in high-visibility areas of the facility that show actual energy and water usage. They should be tied directly to the sub-meters or the building-management system. This approach allows building occupants the ability to see how their habits are contributing to, or hindering, water conservation.

Section 10.0

Waste Management

10.1 General

Minimizing and managing waste can bring significant benefits to the owner/operator and the environment. Waste disposal costs continue to increase as landfill space becomes scarcer and as standards of disposal are tightened.

The Waste Management Plan (WMP) should serve as a roadmap for the facility owner/operator to achieve their overall sustainability goals. The WMP identifies strategies to collaborate with the residents, businesses, commercial haulers, and other stakeholders to reduce waste across all sectors and ensure that all materials are managed for their highest and best use to minimize environmental impacts. The WMP will seek to meet regulations and goals established by the Environmental Protection Agency, local and state regulations while focusing on strategies that address the Building facility's unique systems and needs.

Figure 10.1 Solid Waste Management Hierarchy

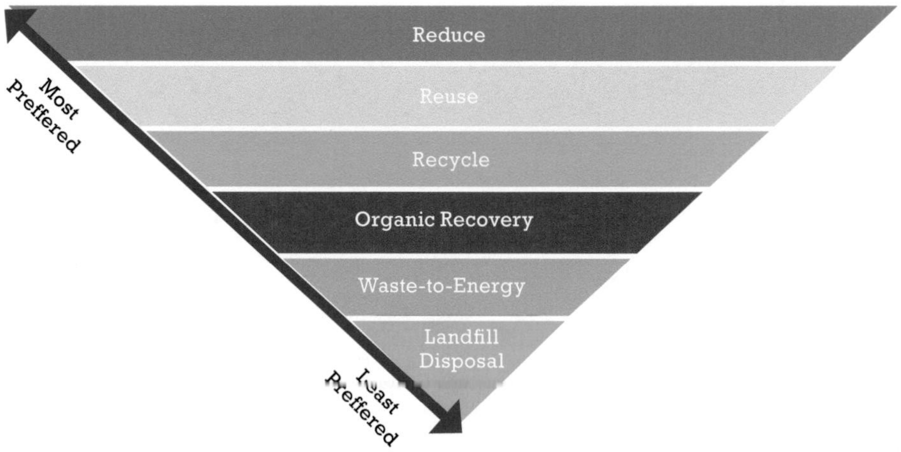

The Plan strategies should be developed using the policy framework reflected below in the waste management hierarchy which characterizes industry best management practices.

The EPA describes the Sustainable Materials Management program as a systematic approach to using and reusing materials more productively over their entire lifecycles. The approach seeks to:

- Use materials in the most productive way with an emphasis on using less.
- Reduce toxic chemicals and environmental impacts throughout the material lifecycle.
- Assure we have sufficient resources to meet today's needs and those of the future.

10.2 Definitions

There are many terms, which relate to the types and sources of wastes and these must be defined. Based on the source, origin and type of waste a comprehensive classification is described below:

a) Domestic/Residential Waste:

This category of waste comprises the solid wastes that originate from single and multi-family household units. These wastes are generated as a consequence of household activities such as cooking, cleaning, repairs, hobbies, redecoration, empty containers, packaging, clothing, old books, writing/ new paper, and old furnishings. Households also discard bulky wastes such as furniture and large appliances which cannot be repaired and used.

b) Commercial Waste:

Included in this category are solid wastes that originate in offices, wholesale and retail stores, restaurants, hotels and other commercial establishments. Some of these wastes are further classified as garbage and others as rubbish.

c) Garbage:

Garbage is the term applied to animal and vegetable wastes resulting from the handling, storage, sale, preparation, cooking and serving of food. Such wastes contain putrescible organic matter, which produces strong odors and therefore attracts rats, flies and other vermin. It requires immediate attention in its storage, handling and disposal.

d) Rubbish:

Rubbish is a general term applied to solid wastes originating in households, commercial establishments and institutions, excluding garbage and ashes.

e) Bulky Wastes:

In this category are bulky wastes which cannot be accommodated in the normal storage containers. For this reason, they require special collection. Bulky wastes are large appliances such as cookers, refrigerators and washing machines as well as furniture, crates, vehicle parts, tires, wood, trees and branches. Metallic bulky wastes are sold as scrap metal, but some portion is disposed of at sanitary landfills.

f) Construction and Demolition Wastes:

Construction and demolition wastes are the waste materials generated by the construction, refurbishment, repair and demolition of residential apartments, commercial buildings and other structures. It mainly consists of earth, stones, concrete, bricks, lumber, roofing materials, plumbing materials, heating/cooling systems and electrical wires, but when generated in large amounts at building and demolition sites, it is generally removed by contractors for filling low lying areas and by urban local bodies for disposal at landfills.

g) Hazardous Wastes:

Hazardous wastes are wastes of industrial, institutional or consumer origin which, because of their physical, chemical or biological characteristics are potentially dangerous to human and the environment. In some cases, although the active agents may be liquid or gaseous, they are classified as solid wastes because they are confined in solid containers. Typical examples are: solvents, paints and pesticides. Good management practice should ensure that hazardous wastes are stored, collected, transported and disposed of separately, preferably after suitable treatment to render them innocuous.

h) E – Waste

In general, computer equipment is a complicated assembly of several types of materials, many of which are highly toxic, such as chlorinated and brominated substances, toxic gases, toxic metals, biologically active materials, acids, plastics and plastic additives.

E-waste consists of all waste from electronic and electrical appliances which have reached their end- of- life period or are no longer fit for their original intended use and are destined for recovery, recycling or disposal. It includes computer and its accessories- monitors, printers, keyboards, central processing units; typewriters, mobile phones and chargers, remotes, compact discs, headphones, batteries, LCD/Plasma TVs, air conditioners, refrigerators and other household appliances.

Electronic equipment's, especially computers, are often discarded by the households and small businesses not because they are broken but simply because new technology has rendered them obsolete and undesirable. Sometimes, the new software is incompatible with the older hardware leaving customers with no option but to buy new ones.

Some governments are taking the lead on reducing E-waste from electronic products by making producers responsible for taking back their products. This is known as Extended Producer Responsibility (EPR). EPR is a policy approach that requires manufacturers to finance the costs of recycling or safely disposing of products consumers no longer want. The objective of implementing directives is to require manufacturers to improve the design of their products in order to avoid the generation of waste and to facilitate the recovery and disposal of electronic scrap. This can be achieved through the phase out of hazardous materials, as well as the development of efficient systems of collection, re-use and recycling. The ultimate aim is to close the loop of the product life cycle so that producers, who manufacturer the product in the first place and who are ultimately in charge of designing the product, get their products back and assume full responsibility for life cycle costs.

i) Sewage Waste:

The solid by-products of sewage treatment are classified as sewage wastes. They are mostly organic and derive from the treatment of organic sludge from both the raw and treated sewage. The inorganic fraction of raw sewage such as grit is separated at the preliminary stage of treatment, but because it entrains putrescible organic matter which may contain pathogens, must be buried/disposed of without delay. The bulk of treated, dewatered sludge is useful as a soil conditioner but invariably its use for this purpose is uneconomical. The solid sludge therefore enters the stream of municipal wastes unless special arrangements are made for its disposal.

10.3 Physical Characteristics

a) Density

A knowledge of the density of a waste i.e. its mass per unit volume (lbs/ft^3) is essential for the design of all elements of the solid waste management system. For example, considerable benefit is derived through the use of compaction, because the waste is typically of low density. A reduction of volume of 75% is frequently achieved with normal compaction equipment. This would mean that the waste removal vehicle would haul four times the weight of waste in the compacted state than when the waste is uncompacted.

b) Moisture Content

Moisture content of solid wastes is usually expressed as the weight of moisture per unit weight of wet material. A typical range of moisture contents is 20 – 45% representing the extremes of wastes in an arid climate and in the wet season of a region having large precipitation. Values greater than 45% are however not uncommon. Moisture increases the weight of solid waste and therefore the cost of collection and transport. Consequently, waste should be insulated from rainfall or other extraneous water.

Moisture content is a critical determinant in the economic feasibility of waste treatment and processing methods by incineration since energy (e.g. heat) must be supplied for evaporation of water and in raising the temperature of the water vapor.

Characteristics of different types of waste vary as follows:

Type of Waste	Principal Components	Approx. Composition % by weight	Moisture Content %
Rubbish	Combustible waste, paper, cartons, rags, floor sweepings	Rubbish 100% Garbage 20%	25%
Refuse	Rubbish and Garbage	Rubbish 50%	50%
Garbage	Animal & Vegetable Wastes	Garbage 100% Rubbish - to 35%	70%
Animal Solids,Organic Wastes	Carcasses, organs,solid organic wastes	100% Animal Tissue	85%

Weights for Refuse Types	Approx, lbs/ft^3
Garbage Commercial - 75% wet, 25% dry	45
Garbage (Average - 65% wet, 35% dry)	35
Dry Rubbish Misc. from office buildings etc.	7

10.4 Methods of Waste Management

Waste Management flows in a cycle: Monitoring, Collection, Transportation, Processing, Disposal / Recycle. Through these steps facility management can effectively and responsibly manage waste output and their positive effect they have on the environment.

Figure 10.2 Process Map

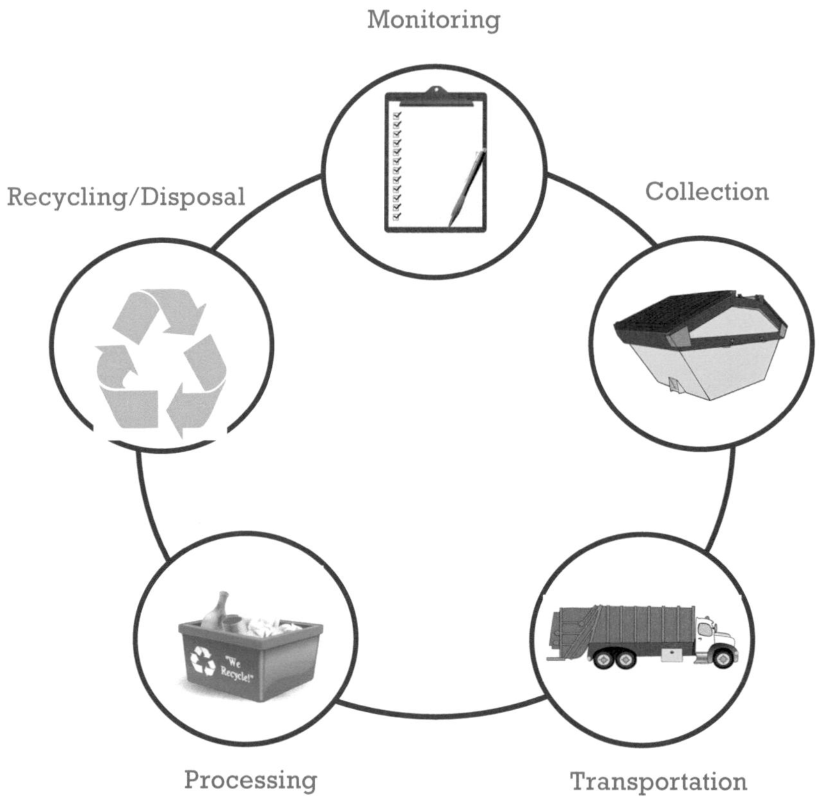

Monitoring is identifying the waste management needs, identifying recycling opportunities and ways to minimize waste output, and reviewing how waste minimization is progressing. Keeping records of the different waste streams is essential to monitor the success of the program.

Collection involves the logistical organization to guarantee that bin containers will not overfill, and waste sit time does not become too long. The correct bin container size and service frequency is a must to prevent overspill or excessive smell.

The correct bins for different wastes must be available with sticker and bin colour identification. Locks, chains, lids and bars prevent public access and non-trained personnel putting rubbish in the incorrect bins.

Cooperation between the waste company and facility management is vital. Bins must be accessible to the truck driver at the agreed times.

Transportation is the organizing of waste transport vehicles with the authorization and ability to transport the specified wastes from the building facility to landfill or processing plant. Vehicles, drivers, and companies need licenses and approval in certain Council Areas to transport waste. EPA standards need to be upheld as well as General Public Safety. Safety standards are vital to the transportation of hazardous wastes. Drivers must undergo training for emergency circumstances that may arise.

Processing involves the separation of recyclables for treatment, and then after treatment are packaged as raw materials. These raw materials are sent to factories for production. Non-recyclable wastes by-pass this step and are delivered straight to landfill. Liquid and hazardous wastes are delivered to treatment plants to become less hazardous to the public and environment.

Disposal / Recycling is the disposal of non-recyclables into landfill. Landfill sites must be approved by legal authorities. Legal authorities guarantee that specific wastes are buried at the correct depth to avoid hazardous chemicals entering the soil, water tables, water systems, air, and pipe systems.

In this step the raw materials made from recyclables are produced and sold as products on the market.

Processing Techniques

I. Mechanical volume reduction

- Compaction
- Balling of papers for recycling
- Used to increase life of landfills

II. Thermal volume reduction

- Incineration

  A waste-to-energy plant (WtE) is a modern term for an incinerator that burns wastes in high-efficiency furnace/boilers to produce steam and/or electricity and incorporates modern air pollution control systems and continuous emissions monitors. This type of incinerator is sometimes called an energy-from-waste (EfW) facility.

III. Manual component separation

- Can be accomplished at the source

Resource Recovery

- Collection, extraction or sorting of recyclable materials from the waste stream
- Recyclable items are bought by manufacturing plants for processing to produce goods

Disposal

IV. Burial

Disposing of waste in a landfill involves burying waste to dispose it of. A properly-designed and well-managed landfill can be a hygienic and relatively inexpensive method of disposing of waste materials.

In a modern landfill, refuse is spread in thin layers, each of which is compacted by mechanical means before the next is spread. When about 10 ft of refuse has been laid down, it is covered by a thin layer of clean earth, which also is compacted. Pollution of surface and groundwater is minimized by lining and contouring the fill, compacting and planting the cover, selecting proper soil, diverting upland drainage, and placing wastes in sites not subject to flooding or high groundwater levels.

V. Composting

Composting is a form of recycling. Like other recycling effort, the composting of yard trimming, food scraps and paper products can help decrease the amount of solid waste that must be sent to a landfill or combustor, thereby reducing disposal costs. These materials are put through a composting and/or digestion system to control the biological process to decompose the organic matter and kill pathogens. The resulting stabilized organic material is then recycled as mulch or compost for agricultural or landscaping purposes. The compost product improves the condition of soil, reduces erosion, and help suppress plant diseases.

There are a large variety of composting and digestion methods and technologies, varying in complexity from simple windrow composting of shredded plant material, to automated enclosed-vessel digestion of mixed domestic waste. These methods of biological decomposition is differentiated as being aerobic in composting methods or anaerobic in digestion methods, although hybrids of the two methods also exist.

Different technologies are used to compost yard trimmings and solid waste feedstocks. These can range simple, low-technology systems that require minimal attention and maintenance to complex systems that use sophisticated machinery and require daily monitoring and adjustment.

10.5 Construction Waste Management

Materials use -- and materials reuse, reduction and recycling -- begins in the planning stages of a project. It starts with the architect, proceeds through the engineer, the estimator, the purchaser, the construction manager and finally the contractors.

Successful materials and waste management – like any successful project -- relies on the skills of many professionals from the architect and designers through project management to the trade contractors.

Table 10.1 presents a list of those people typically involved before the materials arrive on site, and the roles they play in effective materials management.

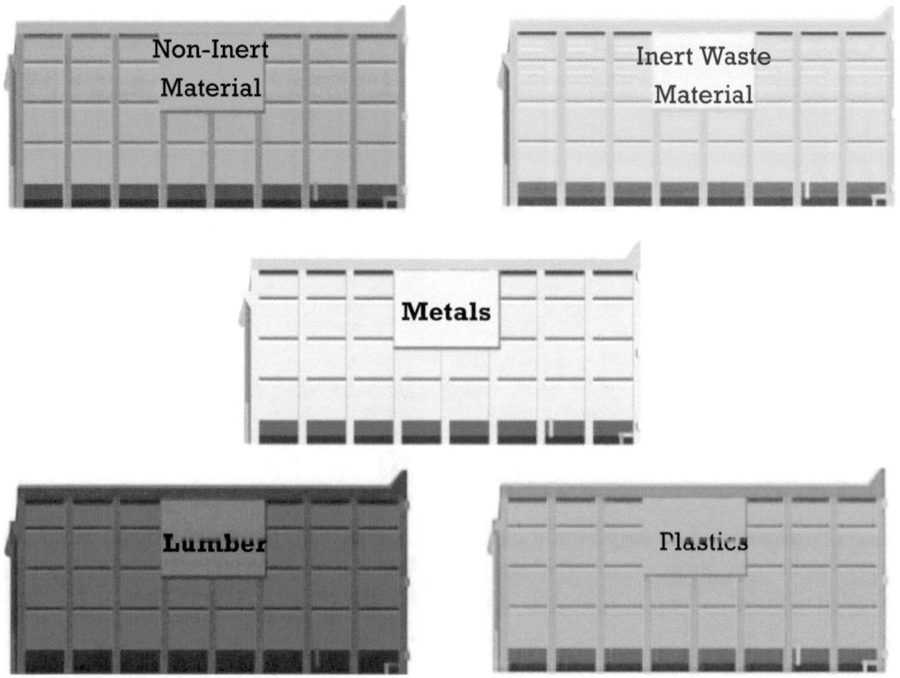

Table 10. I Pre-construction people Involved In Materials Management

Team Member	Role in Materials Management
Architect	Designs for best use of standard sizes, for multiple applications and for their recyclability. Specifies materials with recycled content, responsible packaging and from renewable resources
Engineer	Ensures appropriate structural component dimensions, quality and spacing for use of standard fasteners and materials for multiple applications and recyclability. Specifies materials with recycled content, responsible packaging and from renewable resources.
Quantity Surveyor	Uses latest materials takeoff technologies and exercises accuracy in estimates. Reviews actual waste generation data and updates actual waste factors regularly.
Owner/Operator	Plans purchases and deliveries to reduce surplus and to balance materials maintenance during on-site storage versus transportation energy consumption. Specifies recyclable andreturnable packaging.

Table 10.2 lists those people directly involved in the use of the materials. They should participate in the planning

process, since they know first-hand the actual site and working conditions.

Table10.2 Construction site people involved In Materials Management

Team Member	Role in Materials Management
Site Construction Management	Applies the materials management plan to the site and oversees its implementation. Takes into consideration physical space available and ensures subcontractors are familiar with and committed to the plan
Site Materials Manager	Keeps track of new materials, cuts and used materials; organizes and stores them for availability by the various trades throughout the project in accordance with the materials management plan.
Subcontractors	Communicates with site management and Materials Manager regarding the types of materials they may be able to use for various purposes, even if temporarily. Ensures trades follow the Plan's practices.
Construction Workers	Use materials properly, store new materials properly, handle and cut them carefully for maximum use and minimum waste. Consider using cuts before new pieces.

Table 10.3 lists those people not employed by the owner, contractor or subcontractor are a part of the materials management team as well. Suppliers, haulers and recyclers each play a role in the successful reduction of waste and optimum utilization of materials.

Table 10.3 Off-site people involved In construction Materials Management

Team Member	Role in Materials Management
Suppliers	Use recyclable packaging and returnable containers and pallets, and accept the returned containers and pallets or informs the site ofothers who will accept them.
Recyclers & Haulers	Provide containers for convenient materials storage and retrieval if appropriate. Instruct the site personnel in separation and quality requirements.

One of the goals of green building construction is to minimize construction waste. The best way to accomplish this goal is to first eliminate waste wherever possible by ordering only materials that are necessary, minimizing packing and shipping materials, using standard-size building products wherever possible, and custom-fabricating products where standard-size building products cannot be used. However, it is not possible to eliminate all construction waste, and the contractor needs to develop and implement a waste management plan at the construction site aimed at reducing the amount of waste that is returned to the ecosystem via landfills and incineration. As shown in Figure 10.1, the goal is to recycle or reuse as much construction waste as possible either on-site or off-site.

The contractor needs to first establish a measurable goal for the reduction of construction waste to be disposed. In total, from almost any construction work site, 90% to 95% of all waste materials can be recycled.

Third-party green building rating systems typically award credit toward building certification or verification as a green building based on the amount of waste that is diverted from landfills and incineration using either percent volume or weight as the measurement. If a third-party green building rating system is being used on a building project, the contractor should understand how the rating system accounts for waste diversion and what the percent thresholds are for getting credit.

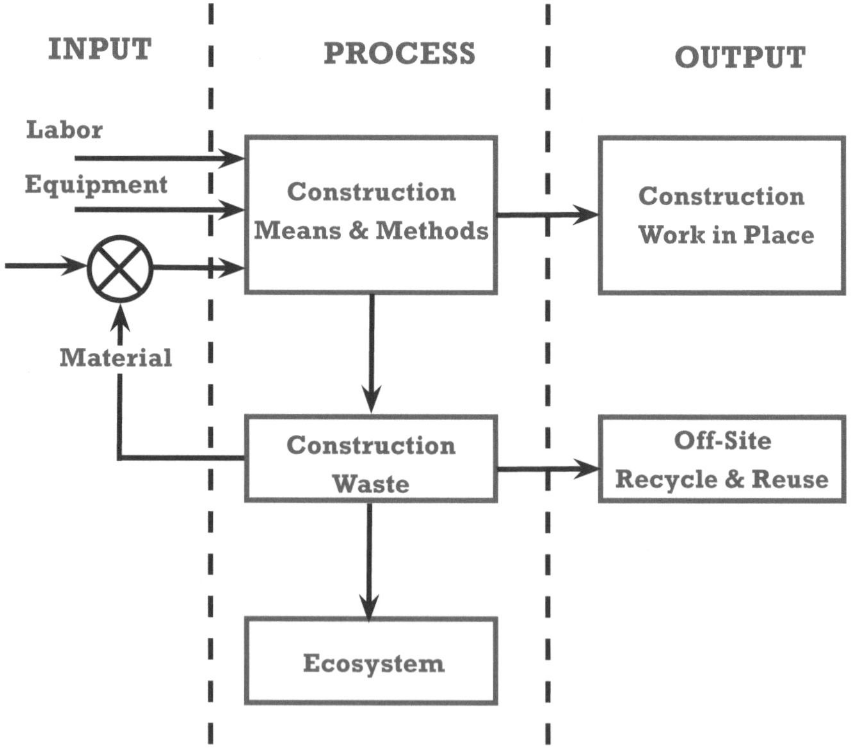

Working with subcontractors, the contractor can then determine what types of waste are expected on the project and a realistic percentage of waste that can be recycled or reused. The percent of project waste that can be recycled on the project depends not only on the amount of waste that the contractor believes will be generated and can be captured at the site, but also what recycling capabilities in terms of both materials and volumes that local waste management companies can handle. Therefore, in addition to working with subcontractors to set a realistic goal for construction waste reduction, the contractor also needs to work with local waste management firms to determine their capabilities to accept the waste.

Once a realistic goal has been set, an on-site waste management plan needs to be developed. This waste management plan should be developed with input from both subcontractors and the local waste management company or companies that the contractor decides to contract with. For a large project with significant volumes of waste, the contractor may want to work with several waste management firms, particularly if they specialize or have capabilities for handling specific types of construction waste. For a small project, there will probably be only one waste management company.

The on-site waste management plan will largely depend the capabilities and services that will be provided by the waste management firm. In most cases, waste will need to be segregated at the project site before pickup, which means multiple bins or Dumpsters to accommodate different materials. If it is not possible to have multiple bins or Dumpsters in a particular area or on-site, then consideration should be given to phased waste management, where bins or Dumpsters are used to collect recyclable waste from construction operations going on at the time and, when complete, the bins are reassigned for waste from the next activity. Collection is not as efficient with phased waste management, because other recyclable waste may not be able to be captured. With proper planning, most of the recyclable waste being generated at any one time should be able to be captured.

In order to ensure success, subcontractors and their field personnel need to understand the importance of waste management to the success of the green building project and what is expected of them. The contractor needs to communicate the project waste management plan to subcontractors, along with the procedures for recycling and reusing construction waste. Wherever possible, the contractor should make it as easy as possible for subcontractors and their field personnel to comply with the waste management plan. This includes placing signs and posters around the site to remind workers to use the recycling bins and Dumpsters instead of just throwing construction waste into general trash Dumpsters. If there are separate bins or Dumpsters on-site for different types of recyclable materials, make it easier for workers to know which is which by using easy-to-read bilingual signs where appropriate, color-coding Dumpsters, or providing another means of identification. Also locate bins and Dumpsters as close to the work area as possible to reduce the distance and time it takes for workers to dispose of recyclable waste properly.

Section 11.0

Products, Materials and Resources

11.1 Green Building Product Characteristics

The term green building materials refers to a growing list of products and materials currently on the market and used to build and furnish buildings. To make these lists, materials have to meet certain eco-friendly criteria such as being manufactured from recycled materials or containing low-Volatile Organic Compound (VOC) levels. The more criteria a product meets, the greener it is considered to be. Selecting environmentally preferable materials requires research, critical evaluation, and common sense. Issues such as code compliance, warranties, and the performance of green products, particularly newly introduced materials, need to be carefully considered. Verification that a product meets advertised claims should be using evaluations based on recognized testing procedures and testing laboratories.

The ideal building material would have no negative environmental impacts. Green material directories are essentially databases that provide listings of available products with the environmental attributes claimed by the manufacturers. It is fairly easy today for a designer or builder to go online and find almost any material.

The concept of producing green materials from recycled materials is very effective and puts waste to good use. Building materials are also classified as green because they can be recycled once their useful lifespan is over. If that product itself is made from recycled materials, the benefit of recyclability is multiplied. Building materials are also considered green when they are made from renewable resources that are sustainably harvested, such as flooring made from sustainably grown and harvested lumber or bamboo. The term "green" can also apply to building materials that are durable. For example, a durable form of cladding can outlast a less-durable product, resulting in significant savings in energy and materials over the lifetime of a building. Moreover, durable products made from environmentally-friendly materials such as recycled waste offer even greater benefits.

Material -sourcing criteria accounts for the contribution from nearly all architectural products in a project. Some credits limit the introduction of contaminants into indoor environments from a products composition. Others concern responsible product sourcing including recycled content, regional production, use of sustainably harvested wood, and the specification of rapidly renewable raw material.

There are several categories of individuals associated with the construction process who require a more rational approach to the evaluation and selection of building products. They can be identified, and their area of responsibility pinpointed as follows:

- Architect and Engineer - The design professional must evaluate and select building products for use in a project.
- Construction Manager - The construction phase expert must make recommendations on building products to bring a project in at the construction cost estimate.
- Subcontractor - The specialist in a building trade becomes a licensed applicator or exponent of specific products.
- Building Materials Manufacturer - The developer of a product who must provide adequate information in the form of product literature.

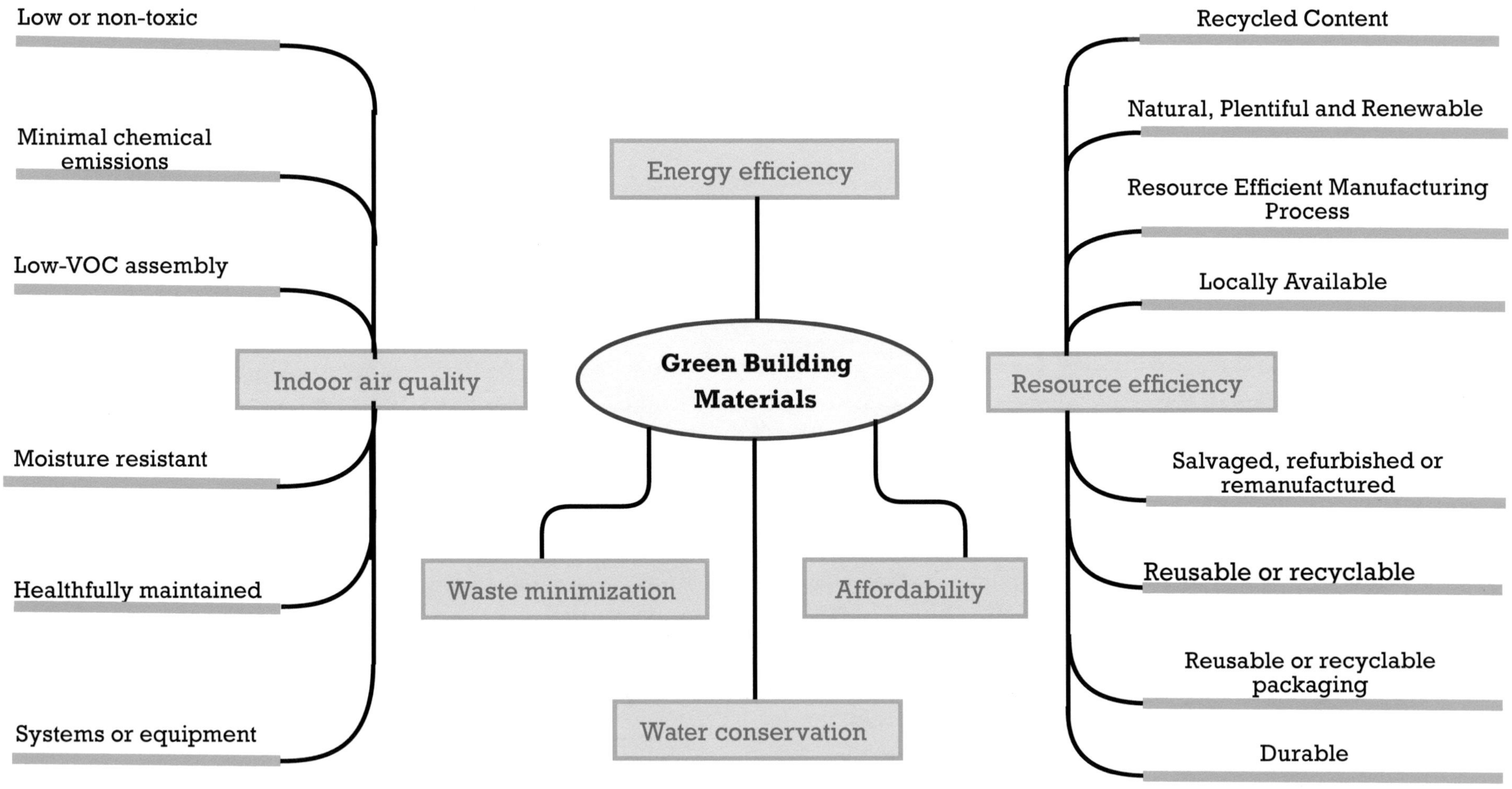
Low or non-toxic
Minimal chemical emissions
Low-VOC assembly
Indoor air quality
Moisture resistant
Healthfully maintained
Systems or equipment
Energy efficiency
Green Building Materials
Waste minimization
Affordability
Water conservation
Recycled Content
Natural, Plentiful and Renewable
Resource Efficient Manufacturing Process
Locally Available
Resource efficiency
Salvaged, refurbished or remanufactured
Reusable or recyclable
Reusable or recyclable packaging
Durable

The evaluation and selection of materials for use in a new project can be separated into two distinct categories:

- Those that are manufactured to meet an existing standard
- Those that are unique in that no standard exists against which they can be measured and therefore require an investigation and assessment as to the claims of the manufacturer.

New materials that fall into the first category are readily evaluated against a product standard. Good product standards are the result of systematic technical research efforts combined with committee work such as ASTM, Federal Specifications, and ANSI standards. These standards establish suitable physical and/ or chemical properties that for the most part have a direct relation to performance. New products that are claimed by the manufacturer as meeting a product standard can therefore be quite readily evaluated against the product standard.

Materials and permanently installed equipment are critical in green building construction and represent a major portion of criteria used to classify or certify a green building. Even though the design team specifies the materials and equipment that will be incorporated into the building, the contractor and its subcontractors must understand the material and equipment specifications as well as the characteristics that make the specified materials and equipment green. This makes material and equipment procurement a critical success factor in any green construction project.

11.2 Building Product Life Cycle

Figure 12.1 illustrates the life cycle of a building product from raw material extraction or harvesting to reuse, recycling, or disposal. Understanding the building product life cycle is very important to understanding the procurement and use of materials and equipment for green building projects. Independently conducted lifecycle assessments (LCAs) and third-party verified environmental product declarations (EPDs) are important tools for comparing some of the environmental attributes of similar products. LCAs and EPDs provide standardized methods for verifying manufacturers' environmental claims and allow for side-by-side product comparisons.

An LCA not only weighs a product's energy efficiency but also measures to what degree its efficiency translates into significant environmental benefits over the full life cycle of the product. It also provides a valuable tool for comparing the product to alternative methods.

11.2.1 Three Phases of Building Materials

Based on the flow of materials, three phases of the building material life cycle can be defined:

- Pre-Building
- Building
- Post-Building

a) The Pre-Building Phase

The Pre-Building Phase describes the production and delivery process of a material up to, but not including, the point of installation. This includes discovering raw materials in nature as well as extracting, manufacturing, packaging and transportation to a building site. This phase has the most potential for causing environmental damage. Understanding environmental impacts in the pre-building phase will lead to the wise selection of building materials. Raw material procurement methods, the manufacturing process itself, and the distance from the manufacturing location to the building site all have environmental consequences. An awareness of the origins of building materials is crucial to an understanding of their collective environmental impact when expressed in the form of a building.

The basic ingredients for building products, whether for concrete walls or roofing membranes, are obtained by mining or harvesting natural resources. The extraction of raw materials, whether from renewable or finite sources, is in itself a source of severe ecological damage. The results of clear cutting forests and strip mining once-pristine landscapes have been well documented.

Mining refers to the extraction, often with great difficulty, of metals and stone from the earth's crust. These materials exist in finite quantities and are not considered renewable. The refining of metals often requires a large volume of rock to yield a relatively small quantity of ore, which further reduces to an even smaller quantity of finished product. Each step in the refining process produces a large amount of toxic waste.

Materials that can be harvested, like wood, are theoretically renewable resources and theoretically more easily obtained without ecological devastation. However, a material is only considered a renewable or sustainable resource if it can be grown at a rate that meets or exceeds the rate of human consumption. Hardwoods, for example, can take up to 80 years to mature.

b) Building Phase

The **Building Phase** refers to a building material's useful life. This begins from the point of its assembly into a structure, includes the maintenance and repair of the material, and extends throughout the life of the material within or as part of the building.

Construction: The material waste generated on a building construction site can be considerable. The selection of building materials for reduced construction waste, and waste that can be recycled, is critical in this phase of the building life cycle.

Use/Maintenance: Long-term exposure to certain building materials may be hazardous to the health of a building's occupants. Even with a growing awareness of the environmental health issues concerning exposure to certain products, there is little emphasis in practice on choosing materials based on their potential for: outgassing hazardous chemicals; requiring frequent maintenance with such chemicals; or, requiring frequent replacements that perpetuate the exposure cycle.

c) Post-Building Phase

The **Post-Building Phase** refers to the building materials when their usefulness in a building has expired. At this point, a material may be reused in its entirety, may have its components recycled back into other products, or it may be discarded.

From the perspective of the designer, perhaps the least considered and least understood phase of the building life cycle occurs when the building or material's useful life has been exhausted. The demolition of buildings and disposal of the resulting waste has a high environmental cost. Degradable materials may produce toxic waste, alone or in combination with other materials. Inert materials consume increasingly scarce landfill space. Adaptive reuse of an existing structure conserves the energy that went into its materials and construction. The energy embodied in the construction of the building itself, and the production of these materials will have effectively been wasted if these "resources" are not properly utilized.

Some building materials may be chosen over others because of their adaptability to new uses. Steel stud framing, for example, is easily reused in interior wall framing, if the building's program should change and interior partitions need to be redesigned. Modular office systems are also popular for this reason. Ceiling and floor systems that provide easy access to electrical and mechanical systems make adapting buildings for new uses quick and cost-effective.

11.3 Specific Requirements

A project's green building product requirements should be clearly defined in the contract documents. Ideally, the green building product requirements will be incorporated into their associated drawings and specification sections. However, green building product requirements are not always explicitly called out in the drawings and specifications. The contractor needs to know what the green building product requirements are for the project and the impact they will have on the contractor's procurement and construction processes. It is very important that the contractor communicate these explicit or implicit requirements to subcontractors, because subcontractors may not be aware of the unique requirements of a green building product.

11.4 General Performance Requirements

Green building product requirements should be incorporated into the contract documents in a variety of ways. For example, the bid documents could require that the project be certified to the silver level using the LEED Green Building Rating System for New Construction and Major Renovations (LEED-NC).

11.5 Green Building Product Characteristics

General green building product characteristics include the following:

- Resource efficiency
- Waste minimization
- Indoor air quality
- Energy efficiency
- Water conservation
- Affordability

The definitions of each of these green building characteristics, as well as the criteria for including products in these categories, can vary greatly based on who is defining them and the purpose, geographic location, and third-party green building rating system, among others. On any project, the contractor needs to look to the contract documents and the third-party green building rating system being used for the specific green product definitions and criteria that apply to that particular project.

11.5.1 Resource Efficiency - can be accomplished by utilizing materials that meet the following criteria:

Figure 11.1 - Life Cycle of a Building Product

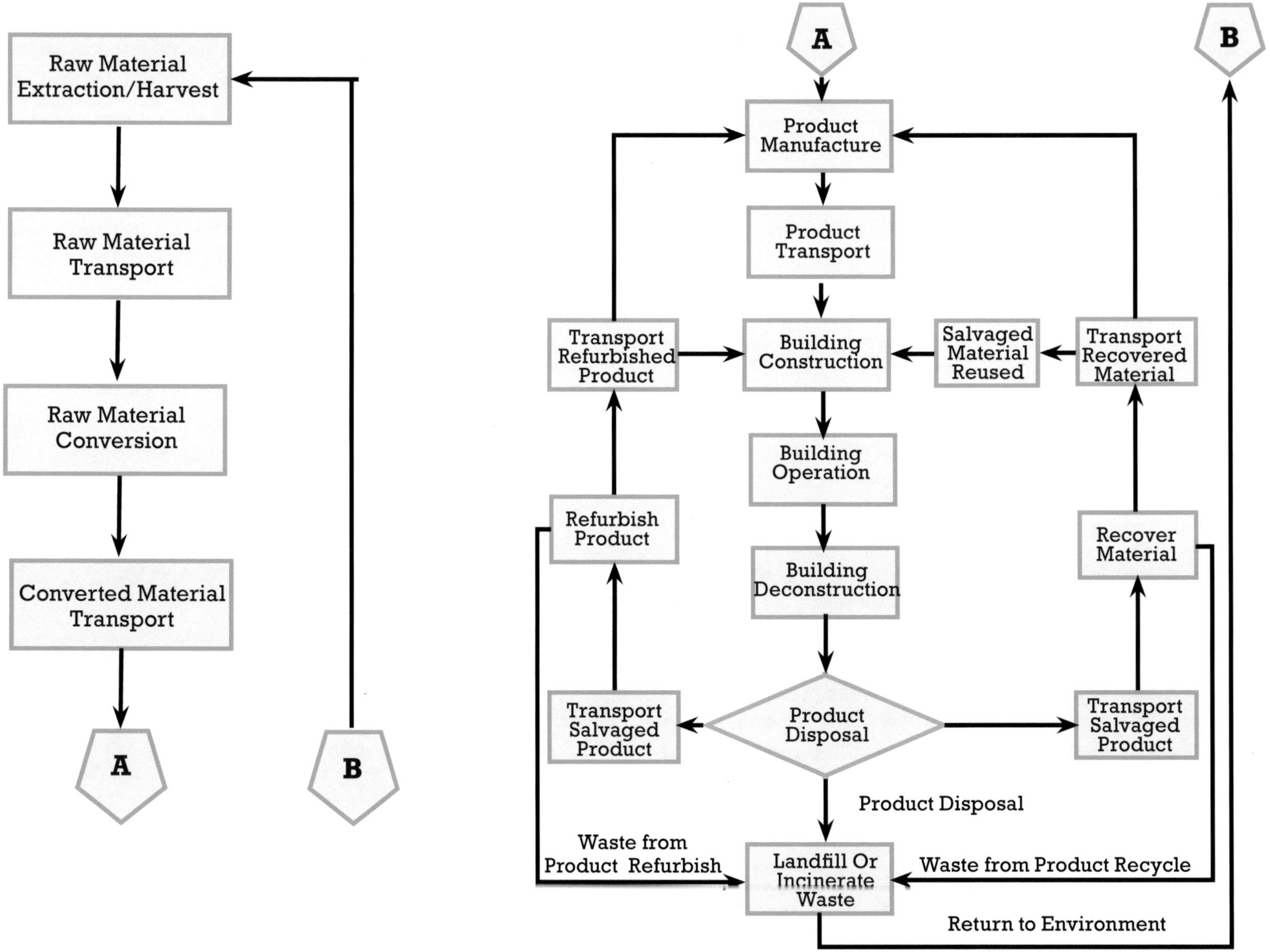

- **Recycled Content**: Products with identifiable recycled content, including postindustrial content with a preference for postconsumer content.
- **Natural, plentiful or renewable**: Materials harvested from sustainably managed sources and preferably have an independent certification (e.g., certified wood) and are certified by an independent third party.
- **Resource efficient manufacturing process**: Products manufactured with resource-efficient processes including reducing energy consumption, minimizing waste (recycled, recyclable and or source reduced product packaging), and reducing greenhouse gases.
- **Locally available**: Building materials, components, and systems found locally or regionally saving energy and resources in transportation to the project site.
- **Salvaged, refurbished, or remanufactured**: Includes saving a material from disposal and renovating, repairing, restoring, or generally improving the appearance, performance, quality, functionality, or value of a product.
- **Reusable or recyclable**: Select materials that can be easily dismantled and reused or recycled at the end of their useful life.
- **Recycled or recyclable product packaging**: Products enclosed in recycled content or recyclable packaging.
- **Durable**: Materials that are longer lasting or are comparable to conventional products with long life expectancies

Waste Minimization strategies could be implemented by the contractor during the building product procurement process, include the following:

- Order only what is actually needed
- Minimize shipping and packing materials
- Use standard size material wherever possible
- Consider custom-fabricated materials
- Consider pre-fabricating material assemblies off-site

The biodegradability of a material refers to its potential to naturally decompose when discarded. Organic materials can return to the earth rapidly, while others, like steel, take a long time. An important consideration is whether the material in question will produce hazardous materials as it decomposes, either alone or in combination with other substances.

11.5.2 Low Emitting Materials

As more and more people become aware of the dangers of global warming and the benefits of sustainable design and construction, the more people are seeking out "healthy" buildings in which to live and work in. For some, the issue is living free of allergies, asthma, and other breathing problems, partly caused by use of materials that emit hazardous substances. As property owners and employers, they invest in air-cleaning systems, buy formaldehyde-free furnishings, and use nontoxic paints and finishes. They are concerned about preventing air pollution problems that may be caused by construction techniques and building materials.

Indoor Air Quality (IAQ) is enhanced by utilizing materials that meet the following criteria:

- **Low or non-toxic:** Materials that emit few or no carcinogens, reproductive toxicants, or irritants as demonstrated by the manufacturer through appropriate testing.
- **Minimal chemical emissions**: Products that have minimal emissions of Volatile Organic Compounds (VOCs). Products that also maximize resource and energy efficiency while reducing chemical emissions.
- **Low-VOC assembly:** Materials installed with minimal VOC-producing compounds, or no-VOC mechanical attachment methods and minimal hazards.
- **Moisture resistant:** Products and systems that resist moisture or inhibit the growth of biological contaminants in buildings.
- **Healthfully maintained**: Materials, components, and systems that require only simple, non-toxic, or low-VOC methods of cleaning.
- **Systems or equipment:** Products that promote healthy IAQ by identifying indoor air pollutants or enhancing the air quality

Energy Efficiency can be maximized by utilizing materials and systems that meet the following criteria:

- Materials, components, and systems that help reduce energy consumption in buildings and facilities.

Water Conservation can be obtained by utilizing materials and systems that meet the following criteria:

- Products and systems that help reduce water consumption in buildings and conserve water in landscaped areas.

11.5.3 Affordability - can be considered when building product life-cycle costs are comparable to conventional materials or as a whole, are within a project-defined percentage of the overall budget.

11.6 Materials

a) Natural versus Synthetic

There are basically two types of building materials used in the construction industry:

- Natural
- Synthetic.

Natural materials are those that are unprocessed or minimally processed by industry, such as timber or glass. Synthetic materials on the other hand are manufactured in industrial settings after undergoing considerable human manipulations, such as plastics and petroleum-based paints. Both have their uses and their advantages and disadvantages. Certain materials and techniques are better suited for specific geographical locations. The choice of material is very important to achieve success.

* Natural Building Materials

They are all around us—mud, stone, and timber are among the most basic natural occurring building materials. And because many of these materials are available throughout the world, the costs and pollution associated with the transportation of these materials across the country decreases. Using natural materials also reduces the amount of toxins in buildings. Additionally, many of these techniques are energy efficient, inexpensive, and easy to build with little required construction knowledge. Some of the popular naturally occurring building materials are:

- Cordwood
- Stone, granite
- Bamboo
- Lumber.

* **Synthetic Materials**

Plastic is a good example of a typical synthetic material. The term plastic covers a range of synthetic or semi-synthetic organic condensation polymerization products that can be molded or extruded into objects, films, or fibers. Plastics vary immensely in heat tolerance, hardness, and resiliency. This adaptability and the general uniformity of composition, as well as the lightness of plastics, facilitates their use in a wide variety of applications. As previously described, there are several criteria that determine a material's greenness, such as whether the material is renewable and resource efficient in its manufacture, installation, use, and disposal. Other criteria include whether the material supports the health and well-being of occupants, construction personnel, the public, and the environment, whether the material is appropriate for the application, and what if any environmental and economic trade-offs exist among alternative materials. There remains a substantial amount of research that needs to be conducted to satisfactorily evaluate the best material alternatives for a project.

Majority of available green building materials and products have one or more of the following properties attributed to them that relate to health and/or environmental issues:

- Durability
- Promote good indoor air quality (e.g. through reduced emissions of VOCs and/or formaldehyde)
- Readily recyclable or reusable when no longer needed
- Can be salvaged for reuse, refurbished, remanufactured, or recycled
- Are locally extracted and processed
- Are salvaged from existing or demolished buildings for reuse
- Less energy is used in extraction, processing, and transport to job site
- Are made using natural and/or renewable resources
- Do not contain ozone depleting substances like CFCs or HCFCs
- Do not contain highly toxic compounds, nor does production result in highly toxic by-products
- Manufactured from waste material such as fly ash or straw
- Are obtained from local resources and manufacturers
- Are biodegradable
- Employ sustainable harvesting practices for wood and bio-based products.

11.6.1 Paints and Coatings

Paints consist of a substance composed of solid coloring matter suspended in a liquid medium and applied as a usually opaque protective or decorative coating to a surface. Primers are basecoats applied to a surface to increase the adhesion of subsequent coats of paint or varnish. Sealers are also basecoats but are applied to a surface to help reduce the absorption of subsequent coats of paint or varnish, or to prevent bleeding through the finish coat. Figure 11.1 reflects the allowable VOC levels stipulated by South Coast Air Quality Management District (SCAQMD).The specifier may obtain the latest information on low VOC products from the Master Painters Institute (MPI).

Figure 11.1 VOC Levels

Paint Type	VOC Limit grams per liter
Flat	50
Non-flat	50
Primers, sealers & undercoats	100
Quick-dry enamels	50

Green paints are paints that are manufactured using Linseed, Soy, Citrus oil, lemon peel oil, natural minerals or other plant oils. The main benefit of using green paints is that less or no toxic chemicals are present in the product, less or no toxic by products are produced in the manufacturing process and the paints are safer to human health as they don't contain lead, formaldehyde, mercury, arsenic, or other harmful chemicals. Another benefit of using green paints is that they tend to outlast and outperform paint made using more toxic materials..

Plant based paints are not water resistant however they do "allow the substrate to 'breathe', are anti-static (avoiding dust), discourage mould growth, and improve air quality and a particular benefit of mineral and clay based natural paints is that they are resistant to cracking, peeling and blisters.. Colour ranges are the same for natural based paints as for traditional paints while lucidity qualities also do not differ from traditional paints.

Paints have significant environmental and health implications in their manufacture, application, and disposal. Most paint, even water-based "latex," is derived from petroleum and creates air pollution and solid/liquid waste. VOCs are typically the pollutants of greatest concern in paints. VOCs emitted from paint and other building materials are associated with eye, lung, and skin irritation, headaches, nausea, respiratory problems, and liver and kidney damage. However, there are renewable alternatives, such as milk paint, that address many of these concerns, although some products are only suitable for indoor applications and at an increased cost. Reformulated low and zero-VOC latex paints with excellent performance in both indoor and outdoor applications.

Paints that meet the Green Seal's GS-11 standard are low in VOCs and aromatic solvents and do not contain heavy metals, formaldehyde, or chlorinated solvents, while meeting stringent performance requirements. Another type of paint that is solvent-free and that may be used on concrete, stone, and stucco is silicate paint. Silicate paint has many advantages being odorless, nontoxic, vapor permeable, naturally resistant to fungi and algae, noncombustible, colorfast, light-reflective, and resistant to acid rain. Additionally, these paints possess extraordinary durability and cannot spall or flake off and will only crack if the substrate cracks.

11.6.2 Coating System

A coating is defined as a liquid, liquefiable, or mastic composition which is converted to a solid protective, decorative, or functional adherent film after application as a thin layer.

Coating generally refers to materials used for protective or functional purposes, so that varnishes and clear coats are designated as coatings. The most successful adaptations to manufacturing low VOC coatings have been two-pack, chemically cured coating systems such as the epoxy, polyurethane, polyester, and vinyl ester systems.

Emission from coating films

For furniture or furniture surfaces the evaporation of odor intensive or even physiologically harmful components of the coating, wood substrate or plastics is relevant. For example, furniture emissions in closed rooms must not exceed specified limits. Also, solvents remaining in the cured coating can constitute a health hazard or add to the pollution level in closed rooms.

The VOC of a coating is determined experimentally using methods defined by the appropriate governing body. The Environmental Protection Agency (EPA) has established Method 24 for VOC determination.

11.6.3 Wood

Native hardwoods - whether oak, maple, cherry or ash - can provide a solution, withe potential to meet requirements of sustainability, aesthetics, durability and ease of maintenance.

Different green rating systems recognize and award points for the use of wood in projects. When awarding points for using wood, these rating systems rely on certification systems to verify the materials come from sustainably managed forest.

There are several types of wood used in building construction including:

a) Pressure Treated Lumber

The popularity over the decades for pressure treated lumber has been partially due to its resistance to rot and insects. But with rapid changes taking place in the treating industry, it is more important than ever to ensure that the treated wood meets standard specifications. There are several arsenic-free preservative formulations on the market, all of which rely heavily on copper as their primary active ingredient.

b) Certified Forest Products

Forest certification was established to help protect forests from destructive logging practices. Where salvaged or reclaimed wood is not available or applicable (i.e., structural applications), specify products that are certified by an approved and accredited organization such as the Forest Stewardship Council (FSC) or the Sustainable Forestry Initiative (SFI). This can also help in achieving LEED or Green Globes certification.

c) Engineered Wood Products

The term engineered lumber, also called composite wood and man-made wood, consists of a range of derivative wood products that are manufactured by pressing or laminating together the strands, particles, fibers, or veneers of wood with a binding agent to form composite materials. It basically refers to a family of engineered wood panels that includes particleboard, medium density fiberboard (MDF), and hardboard. The superior strength and durability of engineered lumber allows it to displace the use of large mature timber. It is also less susceptible to humidity-induced warping than equivalent solid woods, although the majority of particle and fiber-based boards require treatment with a sealant or paint to increase water penetration resistance. These products are engineered to meet precise application-specific design specifications that are tested to meet national or international standards. Engineered wood products also have some disadvantages; for example, they require more primary energy for their manufacture than solid lumber. Furthermore, the adhesives used may be toxic.

11.6.4 Flooring Systems

Few areas in a building will recieve the wear and care as the floor.Consequently design and construction specialists must be familiar with many choices involved selecting a floor covering.

a) Carpet

The manufacture, use, and disposal of carpet have significant environmental and health implications. Most carpet products are synthetic, usually derivatives of non- renewable petroleum products; and its manufacture requires substantial energy and water and creates harmful air and solid/liquid waste. However, many carpets are now available with recycled content, and a growing number of carpet manufacturers are refurbishing, and recycling used carpets into new carpet. Likewise, redesigned carpets, new adhesives, and natural fibers are available that emit low or zero amounts of VOCs. For improved air quality selected carpets and adhesives should meet a third-party standard, such as the Carpet and Rug Institute (CRI) Green Label Plus. The main reasons carpets comprised of natural fibers are often preferred to synthetic carpets is because they are more environmentally friendly, and they have an inherent resistance to dirt, renewable, recyclable and biodegradable. Options include wool, jute, sisal, silk, and cotton floor coverings. Biodegradable carpets made from plant extracts and plant-derived chemicals are also available. Among the disadvantages of carpets is that they harbor more dust, allergens, and contaminants than many other materials. Better indoor air quality can be achieved with the use of other durable flooring materials such as a concrete finish, laminate, vinyl, stone or reclaimed hardwoods to name but a few potential alternatives.

b) Tile Products

Tile is manufactured primarily from fired clay (porcelain and other ceramics), glass, or stone, and provides a useful option for flooring whose principle environmental benefit is durability. One of the principal advantages of tile is its durability as it can last almost indefinitely even in high-traffic areas, eliminating the waste and expense of replacement. The durability of ceramic allows it to be used in areas subjected to heavy loads and high levels of traffic.

Tile production however is energy intensive, although tile from recycled glass requires less energy than tile from virgin materials. Tile's attributes include that it is non- flammable, will not retain liquids, does not absorb fumes, odors, or smoke and, when installed with low- or zero-VOC mortar, can contribute to good indoor air quality. The production process for ceramic glazed tile is the same as for ordinary ceramic tile, except that it includes a step known as glazing. Glazing ceramic tile requires a liquid made from colored dyes and a glass derivative known as flirt that is applied to the tile, either using a high-pressure spray or by direct pouring. This in turn gives a glazed look to the ceramic tile. The most popular types of tiles in use today are glazed and unglazed floor tiles.

When used for flooring, tiled surfaces do not trap dirt and dust, making them easy-to-clean and hygenic flooring solution

c) Vinyl

Polyvinyl chloride's (PVCs) general properties—abrasion resistant, lightweight, good strength relative to its weight, water resistance, and durability—are key technical advantages for its use in building and construction applications. It can also be made scratch resistant, sunlight resistant, and of almost any color. Furthermore, rigid PVC is inherently difficult to ignite and stops burning upon removal of the heat source. Tests show that when compared to its common plastic alternatives, PVC performs better in terms of lower combustibility, flammability, flame propagation, and heat release. Vinyl tends to be inexpensive, in part because vinyl production typically requires roughly half the energy needed to produce other plastics. Products made from vinyl can be resistant to biodegradation and weather and are effective insulators.

However, there are serious environmental concerns regarding the use of vinyl. Vinyl's life-cycle begins and ends with hazards, most stemming from chlorine, its primary component. Chlorine makes PVC more fire resistant than other plastics. Vinyl chloride, the building block of PVC, causes cancer. Lead, cadmium, and other heavy metals are sometimes added to vinyl as stabilizers; and phthalate plasticizers, which give PVC its flexibility, pose potential reproductive risks. Manufacturing vinyl or burning it in incinerators produces dioxins, which are among the most toxic chemicals known to man. Research has shown that the health effects of dioxin, even in minute quantities, include cancer and birth defects. There are many substitutes coming on the market but they may be more expensive or require different maintenance. Nevertheless, there are numerous applications where the substitution of eco-friendly alternatives, particularly for indoor applications where occupants can be directly exposed to off-gassing plasticizers, would clearly be prudent for the sake of occupant health and well-being.

Luxury vinyl tile (LVT) and luxury vinyl plank (LVP) differ from traditional vinyl and vinyl composition tile in various ways, namely in the materials used and composition of each. LVT and LVP products have improved indoor air quality and offer low maintenance needs, decreased lifecycle costs etc.

d) Rubber Flooring

Premium rubber flooring combines high-quality rubber, raw mineral materials extracted from natural deposits, and environmentally compatible color pigments with manufacturing processes that create a single homogenous product free of layers. Together, these materials and processes ensure the safety, durability, surface density, stain resistance, maintenance characterizing floors that contribute to positive user outcomes.

Rubber flooring reduces the negative impact of noise. As the foundation of the built environment, flooring plays a primary role in transferring noise generated by equipment, footsteps created by hard-soled shoes, and cleaning equipment.

Premium rubber is resilient, easing the stress of walking and standing while ensuring comfort underfoot allowing users of facilities such as fitness centers to comfortably use the exercise areas. Additionally, premium rubber flooring does not contain any added antimicrobials, which means the flooring is free of pesticides and chemicals.

In addition to maintenance efficiencies, the elimination of harsh cleaning chemicals, coatings, waxes, and strippers improves IAQ which has a direct impact on health and wellness.

Where functionality is the ultimate goal, such as for athletic flooring, durability, shock absorption and acoustics are very important. Rubber flooring produced from recycled tires, is a high-performance, sustainable material that is often specified for athletic facility floors.

In addition to a long service life, recycled rubber flooring requires little maintenance. The material's density, along with the spring surface, is able to compress and then have a cushioning effect. The extra-thick, unique composite structure also provides durability.

Rubber flooring is available in virgin sheet rubber and recycled rubber, which negotiate various issues of sustainability and cost. Recycled rubber is a material that cuts acoustic output and produces a less noisy space.

e) Wood

Developments in hardwood flooring construction,factory applied finishes and installation techniques have resulted in a whole new generation of hardwood floors that are expressely designed, engineered, and manufactured to provide the warmth and beauty of real hardwood in commercial projects.

Specifiers can select low-maintenance, high-performance hardwood floors specifically intended for applications ranging from retail stores, restaurants and sports floors to corporate conference rooms.

The Wood Flooring Manufacturers Association developed a grading system that describes the appearance of hardwoods according to variables such as color, grain, and markings. The grade is the primary determinant of a floor's appearance after it is installed, sanded, and finished.

11.6.5 Miscellaneous Building Elements

a) Gypsum Wall Board

Also known as drywall or plasterboard, gypsum wall board is manufactured in the United States and Canada to comply with ASTM Specification C 1396. This standard must be met whether the core is made of natural ore or synthetic gypsum. Gypsum wall board is a plaster-based wall finish that is available in a variety of sizes; Due to its ease of installation, familiarity, fire resistance, non-toxicity, and sound attenuation. Gypsum wall board is a benign substance (basically paper-covered calcium sulfate). The main advantages of gypsum board include low cost, ease of installation and finishing, fire-resistance, sound control, and availability. Disadvantages include difficulty in applying it to curved surfaces, and low durability when subject to damage from impact or abrasion. Reclaimed gypsum board can easily be recycled into new gypsum panels that conform to the same quality standards as natural and synthetic gypsum but doing this may not be practical because gypsum is an inexpensive material that can require significant labor to separate and prepare it for recycling.

Laminated noise-reducing gypsum board is used for new construction or renovations over both wood and steel framing as acoustic control methods on interior walls and ceilings. This product has the ability to dampen sound transmission by using an inner polymer layer as a kind of shock absorber that slows board vibrations, dissipating the sound energy into thermal 'energy.

b) Roofing Systems

Factors that affect the choice of a roofing system are durability, waste reduction, and potential liability. Other considerations that will impact the type of roof chosen should include:

- The roof's ability to resist the flow of heat from the roof into the interior, whether through insulation, radiant barriers, or both
- The roof's capacity to reflect sunlight and re-emit surface heat (Cool roofs can reduce cooling loads and urban heat-island effects while providing longer roof life)
- The roof's ability to reduce ambient roof air temperatures through evaporation and shading, as in the case of vegetated green roofs
- Roofs that are recyclable and/or have the capability of being reusable, to reduce waste, pollution, and resource use are preferable.
- Roofs comprised of membranes that do not contain bromine or chlorine are preferable.

Buildings located in urban areas are most likely to be impacted by heat island effect. Cool roofng systems offerer significant benefits in terms of comfort and cost. Reflective cool roofs can reduce roof surface temperatures by up to 100°F. This will in turn lower the cooling energy used by a building. Another benefit of the cool roof is that the lower temperature will cause less expansion and contraction of the roof materials, extending their useful life.

Green Roofing

Vegetated roofing is one of the more significant developments in sustainable building design. Depending on load capabilities and other application-driven requirements, green roofs can be planted with herbs, grasses, flowers, even trees, in an exciting variety of colors, textures, scents, and heights. Emerging new technologies that are helping to promote green building are increasing efforts to make useable space of existing and/or new rooftops to provide additional living space. The key to creating these spaces is to use lightweight and recycled materials and have a plan for storm water management. In this respect traditional drainage systems using pipe and stone are not plausible. Green roof systems are a natural way of providing additional clean air through the transference of CO_2 and oxygen between the plants and vegetation with the atmosphere.

c) Insulated Concrete Forms (ICFs)

At their most basic level, ICFs serve as a forming system for poured concrete walls and consist of stay-in-place formwork for energy-efficient, cast-in-place, reinforced concrete walls. The forms are hollow, lightweight, interlocking modular components that are dry-stacked (without mortar) to create a formwork system into which concrete is poured. The forms lock together and serve to create a form for the structural walls of a building. Concrete is pumped into the cavity to form the structural element of the walls. Insulated concrete forms (ICFs) use an insulating material as permanent formwork that becomes a part of the finished wall. ICFs can be considered "green" materials as they are durable, produce little or no waste during construction, and dramatically improve the thermal performance of concrete walls.

Standard concrete is a dense material with a high heat capacity that can be used as thermal mass, reducing the energy required to maintain comfortable interior temperatures. However, concrete lacks good insulation qualities and standard formwork is waste intensive, so that toxic materials are frequently needed to separate formwork from the hardened product. ICFs address these weaknesses by reducing solid waste, air and water pollution, and potentially reducing construction cost. ICF wall systems have superior thermal qualities that enhance their usefulness for passive heating and cooling; comfort is also enhanced, and energy costs are reduced. ICFs also offer the structural and fire-resistance benefits of reinforced concrete; structural failure due to fire is rare to nonexistent. Due to the addition of flame-retardant additives, polystyrene ICFs tend to melt rather than burn, and interior ICF walls tend to contain fires much better than wood frame walls, thereby improving overall fire safety.

11.6.6 Adhesives, Sealants, and Finishes

Sealants have the ability to increase the resistance of materials to water or other chemical exposure, and caulks and other adhesives help control vibration and strengthen assemblies by spreading loads beyond the immediate vicinity of fasteners. Furthermore, both properties enhance durability of surfaces and structures. Many construction adhesives formulas are hazardous in manufacture and often contain more than 30% volatile petroleum-derived solvents to maintain liquidity until application. This can be harmful to workers who become exposed to toxic solvents, and as the materials continue to outgas during curing the occupants can also be potentially exposed to emissions for extended periods of time. With regard to water-based adhesives, they are available from a number of different manufacturers. Industry tests indicate that these products work as well as or better than solvent-based adhesives, can pass all relevant ASTM and American Plywood Association (APA) performance tests, and are available at comparable costs to common solvent-based adhesives. One of the problems surrounding stains and sealants is that they normally emit potentially toxic VOCs into indoor air. An effective approach to manage this problem is employing materials that do not require additional sealing, such as stone, ceramic and glass tile, and clay plasters. The toxicity, and the air and water pollution generated through the manufacture of chlorinated hydrocarbons such as methylene chloride, emphasizes strongly for using responsible, effective alternatives, such as plant-based, sealant formulations that are non-toxic or have low-toxicity.

General LEED requirements stipulate that all adhesives and sealants used on the interior of the building (defined as inside of the weatherproofing system and applied on-site) should comply with the reference standards described in SCAQMD rule 1168.

11.6.7 Insulation

Insulation is a critical component of any building and especially of green buildings, which are designed and built to minimize environmental impacts. Insulation reduces heat loss when it's cold out and unwanted heat gain when it's warm, thus reducing the need for fossil fuels and other energy inputs with their associated environmental impacts. By reducing energy consumption, insulation also saves money.

Determining what type of insulation to install and how much can be complex from many angles—environmental issues, human health, performance, and building science.

In addition to standard fiberglass, cellulose, polystyrene, and polyisocyanurate insulation, insulation materials made from mineral wool, cementitious foam, radiant foil, cellular glass, vacuum panels, gas-filled panels, wool, recycled cotton, and polyester are available.

The primary environmental considerations that should be investigated when selecting insulation materials are as follows:

- Energy savings
- Raw material acquisition
- Embodied energy and embodied carbon
- Hazardous constituents
- Ozone-depleting substances
- Greenhouse gases and global warming potential
- Halogenated flame retardants
- Chemical byproducts and residuals
- Fiber shedding
- End-of-life issues

It is important to address moisture dynamics, airflow, and how different materials in the assembly—including insulation—interact to deliver the desired performance and durability. Thermal properties, control of air leakage, and moisture management, all interact and should be considered together. Doing so ensures durability of the overall assembly and building—a major environmental benefit that transcends material choice.

11.6.8 Regional Materials

The matter of indigenous materials purchase is an economic issue in addition to an environmental issue. The main intent here is to reduce material transport by increasing demand for building products made within the region. This is achieved by increasing demand for building materials and products that are extracted and manufactured within the designated region. It will help reduce the environmental impacts emanating from transportation (trucking, ship, barge, and rail), in addition to supporting the regional economy. Regional materials should also be carefully considered for achieving credits toward project certification. For example, LEED requires that a minimum of 10% or 20% (based on cost) of total building materials and products be extracted, harvested, recovered, or manufactured regionally within a radius of 500 miles of the project site. Either the default 45% rule or the actual cost of materials purchased may be used. Excluded from the calculations should be all mechanical, electrical, plumbing, and specialty items such as elevator equipment.

Section 12.0

Green Information and Communication Technology

12.1 General

Rapid technical development is have resulted in devices that have become almost indispensable to everyday activity; it is impossible to envisage life without things such as cell phones, digital TV on demand, or computer programs and applications. These technologies have become such an integral part of everyday life.

Even without newly emerging uses of existing technology and of the radical new technologies that are yet to take shape, we are in the grip of one very clearly predictable consequence of technological advancement: the fact that more users are making more use of more devices to do more things. In addition, the case that each of these new activities requires more resources as it becomes more complex.

Green computing refers to supporting business critical computing needs with least possible amount of power or sustainable computing. This is a new paradigm of designing the computer system which considers not only the processing performance but also the energy efficiency.

12.2 Internet Saves Energy and the environment

The Internet holds the potential to increase efficiency in a variety of buildings, including retail, residential, and office buildings. Internet energy efficiency is a very powerful tool for reducing building energy intensity. The Internet is driving a boom in purely home-based work.

A variety of wired and wireless technologies are continuously extending the reach of the Internet. Commerce over the Internet, driven principally by business-to-business transactions, is expected to grow at an even faster pace than the online population.

B2B e-commerce will continue to expand because it is a powerful technology that can improve the efficiency of producing and distributing goods and services. It has the potential to streamline the supply chain, from the point of resource extraction to manufacturing, shipping and residuals management. By automating production, tracking inventory more effectively, making transportation more efficient, and improving cradle to grave management, B2B solutions would result in significant environmental benefits while reducing costs.

The Internet has displaced a wide range of printed materials, from textbooks and corporate brochures to encyclopedias. The technology for displaying text on a computer screen has made electronic books make substantial inroads into the market for printed text, through dedicated user-friendly electronic reading devices, where more and more people opt to read electronic versions of paper documents. With paper one of the most energy- and resource-intensive industries in an economy, the environmental savings are substantial. Including transportation, production and raw materials, the carbon dioxide reductions.

Electronic business applications have considerably reduced paper use, with a related reduction in energy and materials consumed to print, store, transport and dispose of bills, contracts, and other physical records of internal business transactions.

A large majority of companies have redesigned their products so that they can receive information from the Internet and be controlled externally. This technology has the potential to make products far more effective, efficient, safe, and long-lasting, with all the attendant environmental, safety and health benefits.

12.3 Information and Communication Technology (ICT) Impact on Pollution

ICT products are composed of many chemical substances that are toxic for the environment and people. For example, beryllium used in semiconductor chips is dangerous for workers manufacturing this electronic equipment, brominated flame retardant used in electronic equipment causing neurobehavioral effects, Cadmium used in the rechargeable computer batteries is toxic for kidneys and bones, and mercury in the flat screens can affect the brain and central nervous system. The management of these chemical substances during the life cycle of electronic products is essential in the green context and must include informing manufacturers, users, and recyclers about the presence of toxic elements. Especially, in the end-of-life product cycle, electronic wastes including hazardous products pollute soil, atmosphere, water, and the beauty of landscape. Incineration releases heavy metals and ashes into the air.

12.4 Green computing

Green computing refers to supporting business critical computing needs with least possible amount of power or sustainable computing. Greening IT products, applications, services, and practices is an economic and environmental imperative, as well as our social responsibility.

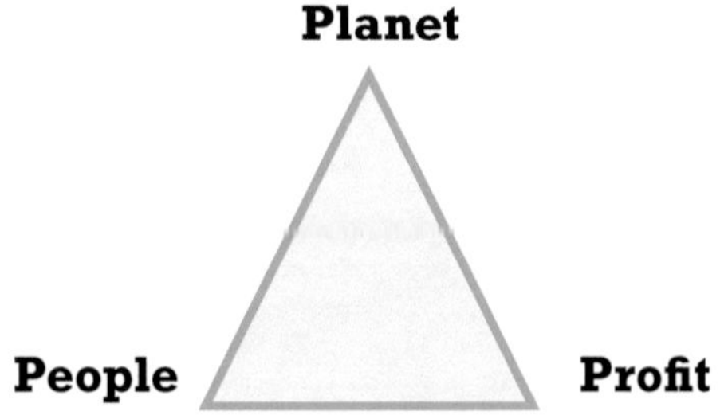

Information Technology (IT) has a large role to play as today's businesses face the need to innovate while keeping the preservation of the eco-system at the core of the innovation. All sustainable development requires a fresh approach to IT and power, putting power consumption at the forefront in all aspects of IT – from basic hardware design to architectural standards, from bolt-on point solutions to bottom-up infrastructure build. The IT function of business is driving an exponential increase in demand for energy, and businesses have to bear the associated costs. The **Triple bottom line (3BL)** consisting of profit, people and planet, has assumed great importance for businesses in order to reduce their carbon footprint.

No longer can businesses merely concentrate on economic performance. They also need to set up a framework to monitor their performance on social and ecological fronts. Eco-innovation implies switching over to a technology that, along with lowering the total cost of operations, also reduces the pressures exerted on the environment. To the ICT manager, it is a means of controlling rapidly increasing energy costs, while also trying to control increasing demands for space and cooling.

12.5 Ethics in ICT

The design of green ICT solutions should integrate all pillars of sustainable development: planet, profit, and people.

Social responsibilities toward the customers: The facility should develop safe and durable ICT products and services, offer efficient customer services, and be prompt, reliable, and courteous in dealing with queries, complaints.

Social responsibilities toward employees: The facility must define a general code of conduct for the management of employees (salary, gender, productivity, etc.) and for the selection of subcontractors according to the consideration of the working conditions of their employees (exploitation of children, no salary, etc.).

A preliminary study should analyze the impact of issues regarding computers in the workplace require assessing the impact of ICT on the elimination of jobs, loss of skills, health issues, and computer crime in installing spyware, hacking and so on.

Social responsibilities to the community: Include educating ICT users to respect the environment, develop good new habits in turning off equipment not being used, turn off mobile phones in meetings, recycle ink cartridges, so forth.

12.6 Environmental Friendly ICT-Products

The impact of the merger of various communication and information services and the Internet can be divided into two very different categories. There are first significant environmental impacts that result from setting up, running and maintaining the ICT equipment that sustains the Internet. A completely different set of environmental impacts will result from the use of Internet applications.

Since there can be no sustainable development without sustainable products, consideration should be given to reduce the environmental strain that results from production, use and disposal of ICT equipment. Environmental impacts such as emissions into the air, the disposal of solid or liquid waste or physical interventions like noise, pressure or radiation are of concern. Parts of the environment that can be effected are the climate system (including the ozone layer), groundwater, plants, animals, soils and many others.

To comprehensively and effectively address the environmental impacts of computing, the company should adopt a holistic approach and make the entire ICT lifecycle greener by addressing environmental sustainability along the following four complementary paths.

Green use — reducing the energy consumption of computers and other information systems as well as using them in an environmentally sound manner

Green disposal — refurbishing and reusing old computers and properly recycling unwanted computers and other electronic equipment

Green design —procuring only environmentally sound components, computers, servers, cooling equipment, and data centers designed on energy efficient principles.

Green manufacturing —manufacturing electronic components, computers, and other associated subsystems with minimal impact on the environment. Strategies which help in implementation of Green Computing Using Virtualization to Reduce Numbers of Servers

Adopting a Green Computing Strategy does not mean a move away from the logical model, only the physical model. Where servers are typically underutilized, virtualization can be used to carve up a single physical machine into a number of virtual servers. From a green perspective the net result is normally a substantial reduced in power and air conditioning requirements saving energy, money and thus reducing the carbon footprint of the server estate.

Virtualization of the desktop mean replacing Personal Computers (PCs) with dumb terminals and migrating to virtual PCs running on a server estate. The user's PC will follow them around wherever they go, whether they are in the office, working from home or even the other side of the globe. If they have remote access it is relatively simple to enable them to run their virtual PC from anywhere. Boot times can also be dramatically reduced, and the users instantly gain the resilience levels of the company's server platforms lower maintenance and cost of ownership.

The green benefits of changing the desktop lie primarily in reduced power consumption, but also that dumb terminals will not need to be upgraded as often as PCs so purchasing and equipment disposal requirements are reduced. By reducing maintenance requirements and also centralizing maintenance the company is able to reduce traveling engineers and support workers, cutting their carbon footprints in the process.

12.7 Sustainable Software Design

Sustainability is a broad term encompassing a large number of concepts in the area of software engineering. The sustainability of software depends on direct effects, which result from executing programs, as well as indirect effects, which influence the sustainability of the product incorporating the software. Indirect effects are caused, e.g., by planned obsolescence when a device manufacturer stops to provide software updates, rendering the device useless. In order to provide sustainable software, the amount of energy consumed during the operation of a device and the energy used during its production should be evaluated carefully.

Improved sustainability during the use of a device by reducing its energy has effects directly noticeable by its users. Especially for battery-operated portable devices, such as laptop computers and mobile phones, a reduced energy consumption during runtime will result in an extended runtime on battery power.

Another application area that is extremely critical in terms of energy consumption is the sensor network. These networks consist of a large number of small sensor nodes, embedded computing devices that communicate over a wireless network. The sensor nodes are commonly distributed over a large area and are either battery powered or rely on energy-harvest-methods. Exchanging depleted batteries or performing maintenance on malfunctioning energy-harvesting devices In these applications is infeasible because of economical as well as physical constraints. As a consequence, sensor network researchers have invented a large number of methods to improve the runtime of sensor nodes on battery or on limited harvested energy using a number of scheduling approaches.

Most of the software approaches exploit a physical property or a specific hardware functionality in order to reduce the power consumption.

In the context of planned obsolescence, software can be used to actively *enforce* the obsolescence of a piece of hardware. The most prominent examples are printers. Many low-cost inkjet printers employ a chip in the ink cartridges that not only ensures that only original cartridges can be used but also implements a counter that prevents them from being used after a certain threshold (number of pages, time, etc.), even though the cartridge may still contain usable ink or could be re-filled. Although this is an example of software used to actually *decrease* the sustainability of printers, there is also software providing countermeasures by, for example, resetting the chip inside the cartridge. Although this software can definitely help to improve the ecological footprint—as well as the economy of operating the printer.

"Software as a Service" or "SaaS" is a technical term for a computer application that is accessed over the internet instead of being installed on a computer or in a local data center. Saas is a service where people connect to online software applications, and have the advantage of accessing the most recent version of any software program.

12.8 Systems Engineering for Designing Sustainable ICT- Based Architectures

Traditionally ICT systems engineers have designed structured cabling and technology systems with only one of their client's departments in mind - the information technology (IT) group. Now, with lighting, mechanical systems, cellular enhancement and audio-visual equipment coming to rely upon the structured cabling system for connectivity, another department-facilities management-has emerged as an important stakeholder in this aspect of building construction.

As commercial buildings have become more and more intelligent, it's not just phones and computers that have a technical-wiring component, but security and maintenance measures have become technology-driven too. A single-architecture, unified network is the technology that will support tomorrow's applications. With multiple voices at the table, designers are challenged to fully understand each party's needs and develop a unified structured cabling design solution that meets the demands of multiple clients within a client.

Multi-Use buildings that incorporate an apartment complex require triple-play services (television, telephone, internet) to the apartments. An economical state-of-the -art future-ready communications infrastructure that would deliver a strong return on investment is required. A micro duct fiber-optic cabling solution could provide connectivity to each apartment unit cost effectively by removing from the equation much of the labor cost from the installation.

Many of today's businesses rely on advanced, ever-changing technology to increase their agility and flexibility while reducing cost. At the same time, they require mission-critical networks that can deliver resilient, fault-tolerant applications for their employees and customers. Because data centers are a capital investment essential to achieving corporate goals, businesses demand unfailing reliability and fast deployment of these critical spaces while managing complexity and costs. Pre-terminated optical fiber and copper cabling solutions will accomplish a number of these objectives. Pre-terminated cable provides a plug-and-play solution for links between switches, servers, patch panels and zone distribution areas in the building. Pre-terminated cabling could cut installation time by up to 80 percent over field terminations. It reduces downtime with faster, more flexible moves, adds and changes.

Wireless is becoming the dominant network access technology. The increasing breadth of the Internet of Things (IoT) and number of connected devices resulting from it is expansive. Apart from ensuring a future-ready network that can support the direction of technology and applications growth, there are financial advantages to a wireless LAN. The installation of fiber in both the vertical and the horizontal has the potential to lower installation cost.

The development of green ICT solutions is highly complex because they must be based on an analysis of the system as a whole including the effects on business, the environment, and the people in the design loop during its entire life cycle. Therefore, the design of green ICT systems requires formal or semiformal methodologies and tools to specify, verify, and validate measures of performances according to both the user's requirements and sustainable development. Moreover, to reach the global green ICT objectives, the assessment should focus not only on the ICT performance of a system of interest but also on its environmental impact.

12.8.1 Power over Ethernet (PoE) - enabled computing

The advantages of being able to power thin client, desktop, and larger mobile devices has higher-power delivery applications such as emerging IEEE 802.3bt Type 3 60W and Type 4 90W PoE. Today's PoE-enabled computers use 30W Type 2 PoE and 60W Type 3 PoE remote powering technology. This has given rise to a new generation of computers that not only use less energy but also possess greater computing power than before. In addition, LED screen technology continues to improve while using less power. This means that we now have full computers with large and high-definition (HD) screens that are capable of running all workplace applications such as office productivity suites and HD video.

The many benefits of PoE computing are as follows:

- Operation over installation friendly and familiar media
- Safe to use and implement
- More cost effective
- Greener solution
- Increased flexibility
- Redundant power

12.8.2 Blown Fiber Infrastructure

Blown fiber and **jetted fiber** are used to describe the placement of a microfiber cable using compressed air. Not only is this technology process getting the fiber to where they need in order to increase bandwidth to their users, but it is keeping construction costs down by providing a pathway for future expansion.

In order to meet the continuous bandwidth-hungry business applications and ever-increasing network Gbits/sec speeds within a constrained budget microduct and air-jet technology network infrastructure is adopted by most companies. Air-jet technology provides unprecedented benefits for today's high-density, high-speed enterprise networks including the following:

- Virtually unlimited fiber pathway and bandwidth capacity, eliminating congested conduit and duct problems
- Immediately scalable, real-time already futureproofed sustainable network; eliminating dark fiber
- Fast and easy fiber installations, upgrades, and network moves/adds/changes to infrastructure even in hard to reach areas
- Flexibility in infrastructure design, reducing or eliminating the need for innerduct and additional conduit
- No physical disruption to facility grounds or inside building fiber installations, promoting safety
- Continuous fiber runs, eliminating splicing and potential points of network failure

- Faster network restoration, reducing or eliminating network downtime
- More than 90 percent labor cost savings for continuous return on investment

12.9 Sustainable Cloud Computing

Computing has evolved from client-server approach to distribute computing. This has given rise to many innovations and business models with cloud computing being one of them. Cloud computing can be considered an example of green IT. It is now feasible for businesses and end-users to demand storage, processing resources, services, hardware and software on an as-needed basis from the cloud.

The sustainability offered by cloud computing, which is all about moving computing infrastructure online, can be shown by the fact that the specific technology allows small business organizations to access large amounts of computing power in a very short time; as a result, they become more competitive with larger organizations. Third-world countries can also significantly benefit by cloud computing technologies because they can use IT services that they previously could not access due to lack the resources. Cloud computing accelerates the time in the businesses market because it allows quicker access to hardware resources without any up-front investment. In this case, there is no capital expenditure (capex), only operational expenditure (opex). Cloud computing makes possible the realization of new, innovative applications such as real-time, location-, environment-, and context-aware mobile interactive applications; parallel batch processing used for the processing of large amounts of data during very short periods of time.

Cloud computing can be applied to all aspects (social, business, environmental) of sustainability. Specific characteristics of the technology, such as sharing data, allow small businesses to have access to huge computing power and as a result, these businesses can become more competitive. Cloud technology can be adjusted to meet requirements of sustainability. It offers dynamic provisioning of resources, multitenancy (the serving of multiple businesses using the same infrastructure), server use, and the power efficiency of data centers.

The advantages of using cloud technologies are numerous. They provide flexibility by providing IT services, such as software updates and troubleshooting security issues, and allow the development of shared applications, thus promoting online collaboration. Cloud computing does not require the use of high-quality equipment and is easy to use. In addition, it enables the sharing of data between different platforms.

Cloud computing provides new horizons to business competitiveness because it enables small businesses to use new technology in order to compete more efficiently with larger competitors by using IT services more rapidly and in a more secure way. Through cloud computing, companies use the server space they need, therefore reducing the consumption of energy in comparison to that used by on-site servers.

Cloud computing can be considered an example of green IT. Green IT could be defined as technologies that directly or indirectly promote environmental sustainability in organizations. Innovative technological systems therefore must be used in order to achieve sustainability. They can be required to be sustainable by using technologies such as those that manage resources in such a way that performance and power are optimized.

12.9.1 The Three-Ways to Cloud Computing

There are essentially three ways in which a business may replace traditional in-house systems with cloud computing. Specifically, the available options are:

- Software as a Service (SaaS)
- Platform as a Service (PaaS) and
- Infrastructure as a Service (IaaS)

All of the above involve the cloud vendor supplying servers on which their customers store data and run applications. However, the differences between SaaS, PaaS and IaaS concern the level of control that a business has over the applications they use, how these applications are created, and the type of hardware on which their applications are run.

When businesses opt for SaaS they can only run those applications that their cloud supplier has on offer. When they opt for PaaS they can create their own applications but only in a manner determined by their cloud supplier. And when they opt for IaaS they can run any applications they please on cloud hardware of their own choice.

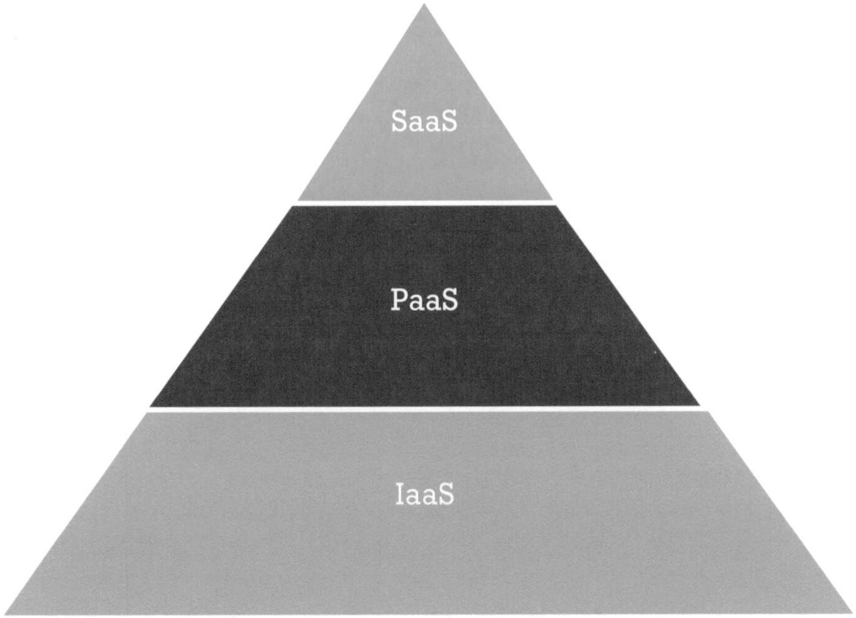

The Building Blocks of Cloud Computing

12.10 Green Audio Visual

AV systems can play a major role in achieving LEED certification in various categories, including the regulation of lighting, the overall reduction of energy use and the "Innovation in Design" category, which addresses sustainable building expertise. Of course, AV-related energy saving initiatives are highly beneficial regardless of whether a institution is actively pursuing LEED

Events are generally huge material and energy consumers and produce large amounts of waste. Increasingly, groups are favoring venues that offer sustainable management practices for their events and meetings. A comprehensive green event offering, at a minimum, includes recycling programs to handle packaging, event materials and food & beverage waste. Event managers also consider green catering options by reducing use of disposable wares, purchasing sustainable food (organic, local, fair trade) and menu planning to reduce waste. Where disposable materials must be used, preference should be given to those with high recycled content and/or other environmentally preferable characteristics (non-toxic inks, biodegradable/compostable, made from rapidly renewable materials, certified sustainably harvested, produced with alternative energy). The flexible nature of room population and use for events requires dynamic designs and systems in order to avoid excessively sized spaces and energy consumption.

There is a need for sustainable AV solutions in meeting rooms, creating solutions that allow users to schedule meetings properly. These solutions enable users find the right space, set the room to their standards, and get it ready for when participants arrive. It also allows the owner to get reports and see how efficiently the rooms are used—what rooms are booked but users are not showing up. Mobile devices are also a key component in sustainable AV, as they help users monitor and control settings in the room to meet green standards.

AV systems can have a significant impact on energy savings. In many commercial settings, the power consumed by electronics not in use – such as lights in an empty room or hibernating monitors – is often the biggest culprit of energy waste and require multiple devices and switches for operation. Solutions, such as room automation technologies, aim to enhance power savings through a single user interface that can control energy output across multiple devices.

12.10.1 Automated Power Systems

AV integrators can help organizations be more energy efficient by using environmental control systems with auto programmable settings to turn off equipment when not in use. Some manufacturers have complete building systems that allow for equipment management and commercial lighting control. Other inventive applications of room automation solutions include end-of-meeting and end-of-day timed shutdowns that serve energy efficiency needs by turning down the power after meetings, workdays or in case of idle activity.

Audio visual systems can have a major impact on company sustainability. AV solutions can play an important role in reducing an organization's carbon footprint. For example:

12.10.2 Business Conferencing

Instead of spending funds on flying employees and partners across the country/globe for meetings, simply have a video conferencing system installed. Not only will the company save money on airfare, car rental and lodging expenses, they will reduce downtime and increase productivity for project teams, not to mention reducing carbon emission related to travel.

Videoconferencing is increasingly popular method for realizing cost and energy savings. By enabling remote telecommuters to work from home or anywhere in the world, videoconferencing reduces travel-related costs and improves a company's environmental footprint. For videoconferencing to truly be transformative, meeting spaces must be equipped to enable remote meetings between groups. While cameras and microphones built into laptops might be suitable for one-on-one conversations, meetings with groups require technologies designed for conference rooms.

With the right tools in place, groups across regions can work together without traveling, encouraging more green and sustainable business practices.

SMART Boards: Meeting organizers can help save trees and dollars spent on supplies for meetings by using interactive SMART Boards and participants remote response systems instead of using paper handouts

Whether it's through consolidation, automation or remote monitoring, businesses today have a wealth of opportunities to utilize today's AV solutions to achieve tangible results across both energy and cost savings.

12.10.3 Remote Monitoring

The continuing alignment between IT solutions and AV equipment and systems has also helped companies achieve meaningful efficiencies. Whereas once AV installations were managed in a disparate manner and no central monitoring, scheduling and management system was in place, today modern software platforms have changed this allow technicians to manage AV assets in real-time. With relevant platforms, necessary personnel can use remote monitoring to power down AV resources that have been running for long periods of time or are in need of going to stand by mode. Such advanced software platforms can unlock capabilities to track and display energy management information for all assets in the system that are being monitored. Technology Systems Managers now have a transparent view of energy utilization that allows them to attain remarkable energy savings over the equipment's lifetime.

12.10.4 Reuse AV products

A very large number of IT products and other electronics are discarded around the world every year. The waste contains both valuable metals and hazardous substances that are often released into nature and affect human health.

Products that are functional should be given a second chance before being collected as waste, and services that enable this are being made visible and used by more people. The prerequisites for the reuse and recycling of technology products continues to be improved.

By following these simple steps, we can all reuse more electronics:

- When buying electronics, check if the product you want is available as a refurbished and upgraded product with a sustainability certification
- Erase old data and sell your products to give them a second life
- If you do not want to sell your old products, hand them into a recycling facility or another collection point

It is not just consumers that can make a difference. By demanding reused products or products that are designed for reuse, large companies can also contribute.

To be able to sell electronics second hand, the products need to be attractive to the buyer and have the potential of functioning well for a reasonable period of time. The rapid development in technology means that products quickly become outdated unless they are leading edge when they are new. Therefore, it is important that the products are high-performing and that hazardous substances are replaced with safer alternatives.

Professional retailers cannot sell products containing substances prohibited by law, even if these substances were allowed when the product was manufactured. Therefore, criteria and certification should be in place so that manufacturers use assessed and approved chemicals whenever possible. Criteria for visual ergonomics are also included for products with a display, partly because this is something that affects how long the product can be used before it needs to be replaced.

The criteria should contribute towards products being functional and attractive for a long time. This will enable both recycling and reuse.

12.10.5 Sustainable AV product purchasing

AV products and their manufacture come with environmental and social risk. By setting and following up on relevant sustainability criteria, purchasing organizations can help drive increased environmental and social responsibility in the market.

Due to the growing worldwide consumption of AV products, and their negative effects, there is an urgency for responsible, sustainable purchasing practices. When organizations demand that AV products are certified for sustainability, this creates incentives for the technology industry to work in a more sustainable direction. Proven results include more environmentally and socially responsible practices from technology brands related to the products and their manufacturing.

Section 13.0

Green Construction

13.1 What is Green Construction

Green construction could be defined as planning and managing a construction project in accordance with the contract documents in order to minimize the impact of the construction process on the environment.

The International Green Construction Code (IgCC), including ASHRAE Standard 189.1 is a regulatory framework that recognizes an entire set of risks not otherwise addressed in the codes. The contractor would bid or negotiate the work in accordance with the contract documents, being mindful of selection criteria that the owner will use to select a contractor for the project. In planning and managing the work, the contractor's project team would look for methods to minimize the impact of the construction process on the environment, which requires:

- Improving the efficiency of the construction process,
- Conserving energy, water, and other resources during construction
- Minimizing the amount of construction waste

Green Construction is a specialized and skilled profession and the contractor should provide strategies that do not adversely impact its project budget or schedule and may even reduce costs and increase productivity. For a particular project, the contractor should look to the specific green building rating system being used for that project to understand its role and responsibilities.

13.2 Green Building Evaluation Systems

Green building evaluation systems are being developed and publicized at the international, national, regional, and local levels. These evaluation systems could be used to guide the design and construction of green buildings.

The U.S. Green Building Council (USGBC) promotes the construction of environmentally friendly, high-performance buildings through its sponsorship of the Leadership in Energy and Environmental Design (LEED) green building rating systems. LEED V4.1 for new construction is a set of rating systems for the design, construction, operation, and maintenance of green buildings including commercial and residential new construction projects. The LEED performance credit system allocates points based on the potential environmental impacts and human benefit of each credit.

Each category has specific design targets worth points. The points awarded are then totalled to earn projects one of four ratings: certified (40-49 points), silver (50-59 points), gold (60-79 points) and platinum (80+ points)

Green Globes is an interactive rating system which is a web-based application for commercial green building assessment protocol offered by the Green Building Initiative (GBI). It offers immediate feedback on the building's strengths and weaknesses and automatically generates links to engineering, design, and product sources. This is a proven method for saving time and money through integrated design and delivery, while benefiting from a cost-effective third-party assessment process. Green Globes NC provides the quickest and most affordable way to assess the environmental sustainability of a new construction project.

Building Research Establishment Environmental Assessment Method (BREEAM) was first launched in 1990 and is updated annually to keep ahead of United Kingdom Building regulations and to stay in line with current best practice.

BREEAM assessment method covers a building performance using the following nine categories:

- Management
- Health & Well Being
- Energy use
- Transport
- Water
- Materials
- Waste Management
- Land use & Ecology
- Pollution

Buildings are rated and certified on a scale of Pass (30-44 points), Good (45-54 points), Very good (55-69 points), Excellent (70-84 points) and Outstanding (85+ points).

Although the foundation for either LEED or BREEAM certification is laid during the design process, the design intent should be implemented through the construction process. The contractor needs to be aware of these requirements, because they can impact material and equipment procurement as well as construction requirements and costs.

13.3 Green Construction Planning and Scheduling

Construction planning and scheduling should account for the unique aspects of green construction. Sustainable project requirements need to be addressed in the construction plan and should be included in the schedule. In particular, green project requirements will impact the procurement, construction, and project close out and commissioning.

Material procurement for sustainable construction projects can impact the sequencing of construction activities as well as activity timing. Project specifications and green building rating systems sometimes require that certain types or a certain percentage of materials be procured regionally or within a given radius around the project site. A large project may strain a regional supplier's ability to support the planned rate of construction. The contractor needs to make sure that local and regional material suppliers are capable of meeting the planned production schedule.

Restrictions on green project site disturbance and concern over contamination of materials stored on the construction site promote just-in-time (JIT) delivery of materials. As a result, the contractor's planning process must account for JIT material delivery, and closer coordination with material suppliers is necessary to ensure an adequate stream of materials to meet planned production rates. It is good practice to include procurement activities in the construction schedule on green projects. Also, the schedule detail associated with material procurement must also increase to provide an effective means of monitoring and controlling material deliveries.

13.4 Elements of Green Construction

Specializing in green building and sustainability basically means incorporating environmentally friendly techniques and sustainable practices into a contracting operation.

13.4.1 Green Design-bid-build Project Delivery

The most traditional construction project delivery system today is the design-bid-build (DBB). It has remained the predominately accepted means by which construction projects are developed and delivered. However, this method is somewhat modified and perhaps more complex with the inclusion of green features into the equation.

With this delivery system, risk is minimized through the owner's control and oversight of both the design and construction phases of the project. The design-bid-build process usually provides the lowest first costs based on submitted tenders but takes the longest time to execute.

The main disadvantages of the traditional design-bid-build process are that it is a lengthy time-consuming process and the owner often has to address disputes that arise between the contractor and design professionals due to mistakes or other special circumstances. With this process the ultimate estimated cost of construction is unknown until bids are finalized, and the system encourages potential change orders. Moreover, there is generally no contractor buy-in to green processes and concepts. An additional challenge with this system is the possibility that construction bids may exceed the owner's stated budget (because plans and specifications are completed prior to tendering the project), the consequence of which is either abandoning the project or redesigning it. Additionally, the owner is required to make a significant financial commitment up front in order to have a complete design in hand as part of the contract documents before solicitation of tenders.

It should be noted that should gaps be discovered between the plans and specifications and the owner's requirements, or errors and omissions are found in the design, it is the owner's responsibility to pay for correcting these errors.

13.4.2 Green Design-build Project Delivery

A distinguishing feature of the design-build approach is that there is only one contract, i.e., the owner contracts with one entity (the designer/builder) that will assume responsibility for the entire project, i.e., its design, supervision, construction, and final delivery. The design-build team is generally led by a contractor (often with a background in engineering or architecture), resulting in the owner issuing a single contract agreement to the contractor, who in turn contracts with a designer for the design.

Many owners prefer the design-build project delivery system to the design-bid-build system because it provides a single point of responsibility for design and construction rather than contracting separately for the design phase and then for the construction with two separate entities. Although it has the advantage of removing the owner from contractor and design disputes, it has the disadvantage of eliminating some of the checks and balances that often occur when the design and construction phase are contracted separately. Additionally, the owner loses much of the project control that exists under a design-bid-build process. The owner also loses the owner/architect advisory relationship that exists in the design-bid-build process.

Having to interact with a single entity has advantages for the owner, such as easier co-ordination and more efficient time management. The design-build contractor or firm will endeavour to streamline the entire design process, construction planning, obtaining permits, etc. Furthermore, with the design-build process activities can easily overlap—construction on parts of the project can begin even while the design is not finalized. At times, the main contractor may involve other organizations on the project with him, but in such cases too, the contractor will be the one dealing with them and assume responsibility. This overlapping offers flexibility to make changes to the design while construction is in process. With the traditional design-bid-build system, this isn't possible, since construction can begin only once the blueprints are finalized.

13.5 Material Conservation

Green procurement is covered in Section 11.0 which focused on the various types and characteristics of green materials that may be incorporated into construction. The contractor needs to be aware of what green materials are, what makes them different from standard building materials, any industry standards that need to be considered when procuring green materials, and special considerations in procuring green materials. The procurement process is very important for the contractor, because it will determine whether the contractor gets materials with the specific green characteristics that it wants or is specified and, if applicable, meets the requirements of the green building rating system being used on the project.

This Section focuses on the green construction process and focuses specifically on waste reduction in the construction process. Material conservation applies not only to green materials but any materials that are brought on site. The material conservation methods discussed in this section will not only reduce the waste stream from the project but also potentially improve productivity, reduce direct construction costs, and improve the contractor's profit.

13.5.1 Material Conservation Planning

Material conservation planning should start with the bid preparation process and continue through the preconstruction process. Material conservation planning should be a team effort, including the designer, if the contractor is involved in the planning and design stages of the project, as well as material manufacturers and suppliers. Where materials will be purchased and installed by subcontractors, these subcontractors should be encouraged to procure their materials and plan their work to minimize material waste, which may require education and training by the contractor. Material conservation planning should be an important part of any constructability reviews or value analyses performed by the construction team during the planning and design process if involved and during bid preparation and preconstruction planning.

13.5.2 Material Conservation Strategies

Some common material conservation strategies are as follows:

- Design to standard material dimensions.
- Fabricate nonstandard material dimensions.
- Prefabricate material assemblies.
- Evaluate material shipping.

The successful application of these and other material strategies will depend on the project characteristics that include the size of the project and the volume of material that is involved.

When considering employing any material conservation strategy, the contractor should consider the overall impact on the project and the project goals regarding waste management and minimization. Reducing the project waste stream to meet specification requirements or fulfill a green building rating system requirement may make the use of a project waste management strategy necessary, even though taken by itself it would not be economical.

a) Build with Standard Material Dimensions

Building with standard material dimensions reduces not only waste but should also reduce material costs through reduced waste and improved field productivity. The ability to build with standard material dimensions should begin in the building planning and design stage. Building dimensions are set during the early stages of design, and it is easiest to consider standard material dimensions at this time. If the contractor is involved with the project during the planning and design, design reviews would be the appropriate time to identify places where changes in building interior or exterior dimensions or shape can facilitate the material installation process. As always, the cost of making a change as well as the impact on other building systems needs to be considered.

If the contractor is not involved in the planning and design of the building, a voluntary alternate included in its bid or a value analysis proposal after contract award may be a way to address the use of standard material dimensions. If the cost savings are large enough or if the proposed change will reduce waste and improve productivity significantly, it may be worth the cost of redesign to the project.

Building to standard material dimensions can be extended to project specialty contractors and suppliers that fabricate materials for incorporation into the work.

b) Fabricate Nonstandard Material Dimensions

If it is not possible to build with standard material dimensions, the next best thing is to fabricate materials to fit the project requirements where the quantity of materials is sufficient to warrant any additional costs that may result from fabricating materials to nonstandard material dimensions. Fabricating materials with nonstandard dimensions requires a careful analysis of the project to determine not only what those dimensions are but also the impact on other related building systems, material shipping to and handling at the project site, and the impact on installation crew productivity, among other considerations.

c) Prefabricate Material Assemblies

Material assemblies could be prefabricated off-site under controlled conditions to reduce waste at the jobsite as well as well as avoid problems that could occur when certain materials are required to be used.

Prefabrication off-site can avoid conflicts between what is needed to meet the contract requirements and provide a quality installation and the requirements of the green building rating systems being used on a project.

d) Evaluate Material Shipping

Waste can occur in shipping as a result of material breakage and spoilage. When planning material deliveries, the contractor should consider the mode of shipping and its impact on material breakage and spoilage as well as the cost of shipping and installation.

13.6 Products and Materials

Materials and permanently installed equipment are critical in green building construction and represent a major portion of criteria used to classify or certify a green building. Even though the design team specifies the materials and equipment that will be incorporated into the building, the contractor and its subcontractors must understand the material and equipment specifications as well as the characteristics that make the specified materials and equipment green. This makes material and equipment procurement a critical success factor in any green construction project.

13.6.1 Building Product Life Cycle

Figure 3.7 illustrates the life cycle of a building product from raw material extraction or harvesting to reuse, recycling, or disposal. Understanding the building product life cycle is very important to understanding the procurement and use of materials and equipment for green building projects.

13.6.2 Three Phases of Building Materials

Based on the flow of materials, three phases of the building material life cycle can be defined:

- Pre-Building
- Building
- Post-Building

a) The Pre-Building Phase

The Pre-Building Phase describes the production and delivery process of a material up to, but not including, the point of installation. This includes discovering raw materials in nature as well as extracting, manufacturing, packaging and transportation to a building site. This phase has the most potential for causing environmental damage. Understanding environmental impacts in the pre-building phase will lead to the wise selection of building materials. Raw material procurement methods, the manufacturing process itself, and the distance from the manufacturing location to the building site all have environmental consequences. An awareness of the origins of building materials is crucial to an understanding of their collective environmental impact when expressed in the form of a building.

The basic ingredients for building products, whether for concrete walls or roofing membranes, are obtained by mining or harvesting natural resources. The extraction of raw materials, whether from renewable or finite sources, is in itself a source of severe ecological damage. The results of clear cutting forests and strip mining once-pristine landscapes have been well documented.

Mining refers to the extraction, often with great difficulty, of metals and stone from the earth's crust. These materials exist in finite quantities and are not considered renewable. The refining of metals often requires a large volume of rock to yield a relatively small quantity of ore, which further reduces to an even smaller quantity of finished product. Each step in the refining process produces a large amount of toxic waste.

Materials that can be harvested, like wood, are theoretically renewable resources and theoretically more easily obtained without ecological devastation. However, a material is only considered a renewable or sustainable resource if it can be grown at a rate that meets or exceeds the rate of human consumption. Hardwoods, for example, can take up to 80 years to mature.

b) Building Phase

The **Building Phase** refers to a building material's useful life. This begins from the point of its assembly into a structure, includes the maintenance and repair of the material, and extends throughout the life of the material within or as part of the building.

Construction: The material waste generated on a building construction site can be considerable. The selection of building materials for reduced construction waste, and waste that can be recycled, is critical in this phase of the building life cycle.

Use/Maintenance: Long-term exposure to certain building materials may be hazardous to the health of a building's occupants. Even with a growing awareness of the environmental health issues concerning exposure to certain products, there is little emphasis in practice on choosing materials based on their potential for: outgassing hazardous chemicals; requiring frequent maintenance with such chemicals; or, requiring frequent replacements that perpetuate the exposure cycle.

c) Post-Building Phase

The **Post-Building Phase** refers to the building materials when their usefulness in a building has expired. At this point, a material may be reused in its entirety, may have its components recycled back into other products, or it may be discarded.

From the perspective of the designer, perhaps the least considered and least understood phase of the building life cycle occurs when the building or material's useful life has been exhausted. The demolition of buildings and disposal of the resulting waste has a high environmental cost. Degradable materials may produce toxic waste, alone or in combination with other materials. Inert materials consume increasingly scarce landfill space. Adaptive reuse of an existing structure conserves the energy that went into its materials and construction. The energy embodied in the construction of the building itself, and the production of these materials will have effectively been wasted if these "resources" are not properly utilized.

Some building materials may be chosen over others because of their adaptability to new uses. Steel stud framing, for example, is easily reused in interior wall framing, if the building's program should change and interior partitions need to be redesigned. Modular office systems are also popular for this reason. Ceiling and floor systems that provide easy access to electrical and mechanical systems make adapting buildings for new uses quick and cost-effective.

13.7 Site Layout and Use

One of the goals of green construction should be to reduce the disturbance to the site during construction in order to preserve the natural setting and habitat around the building. If the site has been previously disturbed, the goal should be to minimize further disturbance and damage to the site and, to the extent possible, return the site to its natural state following construction. Reducing site disturbance not only addresses the building site but also protects surrounding land, both developed and undeveloped, from disturbance or damage during construction. Third-party green building rating systems typically have specific requirements for minimizing site disturbance during construction, and the contractor should be familiar with those specific requirements if credit is sought for minimizing site disturbance. Strategies that the contractor can use to ensure reduced site disturbance include the following:

- Set construction boundaries.
- Restrict vehicle and equipment movement.
- Establish trailer, storage, and laydown areas.
- Prevent site erosion and sediment runoff.
- Manage storm-water and wastewater
- Set Construction Boundaries

In order to prevent site disturbance during construction, the contractor needs to set and mark boundaries around the perimeter of the building, as well as around hard surfaces such as surface parking areas, sidewalks, driveways, and access roads that are part of the construction project. In addition, corridors should be established for incoming utility lines to prevent damage. The boundaries should be clearly marked, and all subcontractors informed of the boundaries within which they will need to work. Subcontractors should be notified of these restrictions before bidding, because these boundaries may restrict the equipment they could use and impact their productivity, which will need to be addressed in their bid.

a) Restrict Vehicle and Equipment Movement

Along with setting construction boundaries, vehicle and equipment movement on the site should be restricted to prevent site disturbance. These boundaries may be the same as those around the building perimeter and hard surfaces, or they may be different for specific equipment to allow its use. If boundaries are extended for specific equipment or construction operations, the contractor should have a plan for restoring the affected area to its natural state. Also, if the contractor is going to make a subcontractor responsible for restoring the site after performing a construction operation outside of the established boundaries, this requirement should be communicated to the subcontractor before bidding and included in the subcontract requirements.

b) Establish Storage, and Laydown Areas

The contractor should formulate and diagram a site layout plan and staging strategy that will provide the highest productivity and efficiency of movement. Storage and laydown areas should be established at the site before mobilization. The contractor should work with subcontractors before mobilization to understand their storage, and laydown area needs throughout the construction process. As a result of the need to minimize site disturbance and limited area with which to work, the contractor may have to restrict the number of trailers that subcontractors are allowed to move on-site as well as restrict storage and laydown areas. Subcontractors may need to move equipment off-site when it is not in use when they would have normally just left it parked on-site until it was needed again. Similarly, storage and laydown areas may be limited and need to be shared by subcontractors as needed during the project.

Once work is complete on a particular phase of a project, excess materials may need to be moved off-site immediately to make room for the next subcontractor's materials. All of this will require the contractor to work closely with subcontractors throughout the construction process to maintain the site and accommodate their subcontractor storage needs for current construction activities

c) Prevent Site Erosion and Sediment Runoff

Erosion is the removal of solids (i.e., sediment, soil, rock, and other particles) by wind, water, or ice in the natural environment. Erosion is always a consideration on a newly graded construction site. The contractor should develop an erosion control plan that should implement measures to prevent site erosion and sediment runoff during construction. The exact measures taken to prevent site erosion and sediment runoff on a particular green building project will depend on soil conditions, site topography, vegetation, and other physical site characteristics as well as contractual, third-party green building rating system, and applicable federal, state, or local government requirements.

This plan may include temporary installations such as silt fences, filter fabrics and straw bale barriers intended to hold soil in place and protect storm sewer inlets. For larger sites, phased grading may be required to limit the extent and exposure of soils to erosion. Disturbed topsoil should be stockpiled and covered to prevent erosion and allow future reuse. Implementing terracing, retaining walls, and re-stabilization techniques prevent long-term erosion

d) Manage Storm-water and Wastewater

Similarly, the contractor should take steps to manage storm-water and wastewater during construction so that runoff does not pollute or damage the building site or surrounding areas

Whenever possible, permanent storm water systems should be designed and implemented to capture and re-use storm-water. Captured water may be used for site irrigation or building use. Before releasing storm-water from the site, low- impact methods of treating the quality of the water and slowing the rate of release should also be examined.

These precautions will prevent the quick release of pollutants from impervious surfaces into nearby water bodies. The contractor should examine landscaping techniques that can be used to reduce impervious surfaces and to alter pollution from storm-water runoff.

13.8 Construction Waste Management

See Section 10.5

13.9 Material Storage and Protection

The contractor should endeavor to schedule material deliveries as close to the actual date of installation of that material as possible. Material storage and protection is very important on green building projects. Materials and equipment, particularly interior finish materials and HVAC and plumbing materials, must be protected to prevent contamination by dust, moisture, dirt, and mold before installation. If possible, materials should be stored off-site in a controlled environment until they are needed for installation. When brought on-site and stored while waiting to be installed, materials should be covered, sealed, and protected from damage and the elements. For instance, fabricated ductwork and piping should have their ends sealed when brought on-site, and the seals should only be removed when the materials are being installed. At the end of the day or if there is a break in an activity, any openings in piping or ductwork should be sealed until work resumes. Porous materials and fabric-based materials should also be kept dry and in a controlled environment until they are ready for installation.

13.10 Providing a Healthy Work Environment

Green construction is most concerned about providing a healthy environment for construction workers and the public during construction. During the demolition and deconstruction of existing structures, earthwork operations, and other outdoor site work, the contractor should control dust and other airborne pollutants as much as possible. For example, dust can be minimized during earthwork operations by keeping the ground damp using water trucks or sprinklers. Similarly, replacing older, heavy equipment with newer, more efficient and cleaner equipment when it is economically justifiable can reduce fuel use and minimize exhaust emissions. This will reduce pollution and improve the environment both on-site for workers and in the surrounding areas for the public.

Once the building is enclosed or nearly enclosed to the point that work can begin inside, the contractor needs to consider the interior environment in which the workers will be working. Of primary concern is the indoor air quality (IAQ), because of the dust that results from many building finish activities, such as drywall and finish carpentry, and the myriad of different chemicals in the form of adhesives, sealants, paints, and coatings that are used in the construction process. The contractor needs to develop and implement an IAQ plan during construction to ensure the health and well-being of workers. In addition, having an IAQ plan that is implemented during construction may be a requirement of the third-party green building rating system.

HVAC protection should address methods to prevent dust from getting into the air-distribution system as well as odors that can be absorbed by porous parts of the HVAC systems and released into the air stream at a later time. The most effective method of controlling pollution is at the source, and the guidelines for source control address methods for containing or eliminating air pollution at the source. If the pollutants cannot be controlled at the source, the contractor should provide recommendations for pathway interruption, which is simply the prevention of air movement along with the migration of dust, contaminants, and odors through the use of barriers and pressurizing spaces.

Housekeeping simply refers to keeping the indoor construction site as clean as possible through dust collection and cleaning. Scheduling involves carrying out construction operations that result in a lot of dust and contaminants at times when the building is not occupied and there is sufficient time for the air to clear before other workers and occupants return.

13.11 Construction Equipment Selection and Operation

The Mixed-Use Building construction industry is of complex work. Projects are big and complicated in nature. In modern green building construction planning process, one of the most important criteria prior to commencement of work would be how equipment can be efficiently used in the construction. With increase in the scope, complexity and nature of work, every single work is executed by a different piece of equipment.

Site conditions-both ground conditions as well as climatic conditions-may affect the equipment-selection decision. Equipment must suit the requirements of work, climate and working conditions.

The contractor should implement several strategies to reduce fuel consumption and pollution resulting from the use of construction equipment and vehicles at the site. These strategies are not only environmentally friendly but will also save the contractor money and increase productivity. Strategies that the contractor might consider to improve the environment, increase productivity, and reduce costs are as follows:

- Select types and quantities of equipment to minimize cycle time,which in turn will increase productivity and reduce fuel costs. The economic considerations such as owning costs, operating labor costs and operating fuel costs of equipment are most important in selection of equipment. The contractor could evaluate economical aspect to decide to rent a piece of equipment that is better suited to the work requirements than to use equipment that the contractor already owns or leases. If there is significant earthwork on a green building project, it may be worthwhile for the contractor to analyze the operation in greater detail than usual and select the type and quantity of equipment that would be best for the job. Equipment dealers often have or have access to simulation software that can assist the contractor select the best equipment spread.

- The contractor should develop guidelines for equipment operators that include a maximum idling time when the equipment will be shut down if the operator thinks that he or she will exceed it.
- The contractor should consider substituting battery-powered vehicles like golf carts that can be recharged overnight for pickup trucks and other vehicles on-site where possible.

13.12 Documenting Green Construction

A green construction project requires that the contractor prepare and provide additional submittals before the start of construction, during construction, and at project closeout. Green project submittals can include plans such as the following that need to be submitted, approved, and documented for compliance:

- Site Preservation and Use Plan
- Waste Management and Recycling Plan
- Indoor Air Quality Plan
- Material Delivery and Protection Plan

Similarly, submittals such as the following that detail the type and quantity of the following material categories may also be required:

- Salvaged and Refurbished Materials
- Recycled Content Materials
- Regional Materials

On green construction projects, product information and certifications in addition to the requirements in the technical specifications should be required. This product information and certification submittal requirement is in addition to the shop drawing process.

The contractor should be advised as to what green construction documentation needs to be supplied. Requirements for green construction documentation will normally be found in the project contract documents and in any third-party green building certification system used on the project.

Section 14.

Green Project Certification and Closeout

14.1 General

Final acceptance and project closeout is initiated upon receiving notice from the contractor(s) that the work or a specific portion thereof is acceptable to the owner and is sufficiently complete, in accordance with contractor documents, to allow occupancy or utilization for the use for which it is intended. Closeout can take place when all contract requirements, warranty, and closeout documents, along with all punch list items, have been resolved.

Documents typically required for project closeout include:

- Signed/sealed as-built or record drawings, which show all changes from the original plans.
- Contractor's certificate of compliance with plans and specifications.
- Architect's issuance of certificate of substantial completion and/or final acceptance and issuing final certificate(s) for payment after a detailed inspection with the owner's representative is conducted for conformity of the work to the contract documents to verify the list submitted by the contractor(s) of items to be completed or corrected.
- Architect's certified copy of the final punch list of itemized work stating that each item has been completed or otherwise resolved for acceptance.
- Determination of the amounts to be withheld until final completion of outstanding punch list items.
- Notification to owner and contractor(s) of deficiencies found in follow-up inspection(s), if any.
- Certification that all closeout requirements, including but not limited to, as-built drawings, warranties, manuals, keys, affidavits, receipts, releases, etc. have been received, reviewed as necessary, and approved for each subcontractor.
- Certificates of use, occupancy, or operation.
- Final waivers of lien in a form satisfactory to the lender and the title company from all subcontractors, suppliers, and the general contractor and indemnifying the owner against such liens.

14.2 Commissioning and Testing

The increased complexity of green buildings means significant time in ensuring proper performance test standards are specified, building codes are satisfied, the design is constructible, and ultimately, Owner's Project Requirements (OPR) is satisfied.

Commissioning and testing of a green construction project is more complex than a traditional building project, particularly if the owner is seeking third-party certification. Green building commissioning should include understanding the contractual requirements for commissioning, the need for a comprehensive mutually agreed-upon commissioning plan early in the project, working with an outside owner-appointed commissioning authority, typical requirements for system start-up and testing, and typical documentation that needs to be submitted. Additionally, typical contract closeout requirements for green buildings should address submission of project documentation such as record drawings, addressing warranties and guarantees, and training the owner's personnel.

The US Green Building Council (USGBC) operates 10

Leadership in Energy and Environmental Design (LEED) Green Building Rating Systems.Eight of those systems - LEED for New Construction (LEED-NC). For LEED-NC, fundamental commissioning is a mandatory prerequisite, and an additional point can be obtained for enhanced commissioning.

14.3 Building Commissioning Purpose and Objectives

Commissioning is a systematic, documented and collaborative process that includes inspection, testing and training conducted to confirm that a building and its component systems are capable of being operated and maintained in conformance with the design intent.

The purpose of building commissioning is to ensure that building systems operate and can be maintained in accordance with the owner's project requirements as expressed in the facility program that defines the owner's operational requirements. The purpose of commissioning is accomplished by achieving the following four commissioning objectives:

I. Verify and document that the equipment constituting building systems to be commissioned has been installed properly and operates correctly based on predefined procedures that include inspection, testing, and start-up.

II. Verify and document that each building system being commissioned operates correctly and interacts as required with other building systems.

III. Ensure that complete equipment and system documentation, including operation and maintenance (O&M) information, is provided to the owner at the end of the project in an organized manner that can be easily accessed and used by the owner's operating personnel.

IV. Ensure that the owner's operating personnel have been properly trained in the operation and maintenance of commissioned building systems in order to ensure reliable, efficient, and sustainable building operation.

With a comprehensive commissioning project manual, the owner should have energy conservation measures identified, operation and maintenance addressed, and operator training fully described.

14.4 The Design Intent Document

A Design Intent document (DID) with a detailed explanation and documentation of the ideas, concepts and criteria that define correct building operation is an essential requirement. The DID should be accompanied by the "basis of design" as described earlier. The basis of design explains how the design team chose certain systems and space arrangements to meet the needs of the occupants.

The intended use of the building leads to the requirements of the DID. It will set the requirements for occupant satisfaction and forms the foundation for the development of the basis of design. The basis of design lists the technical criteria that will be included in the construction documents (i.e., drawings and specifications (see Figure 14.1).

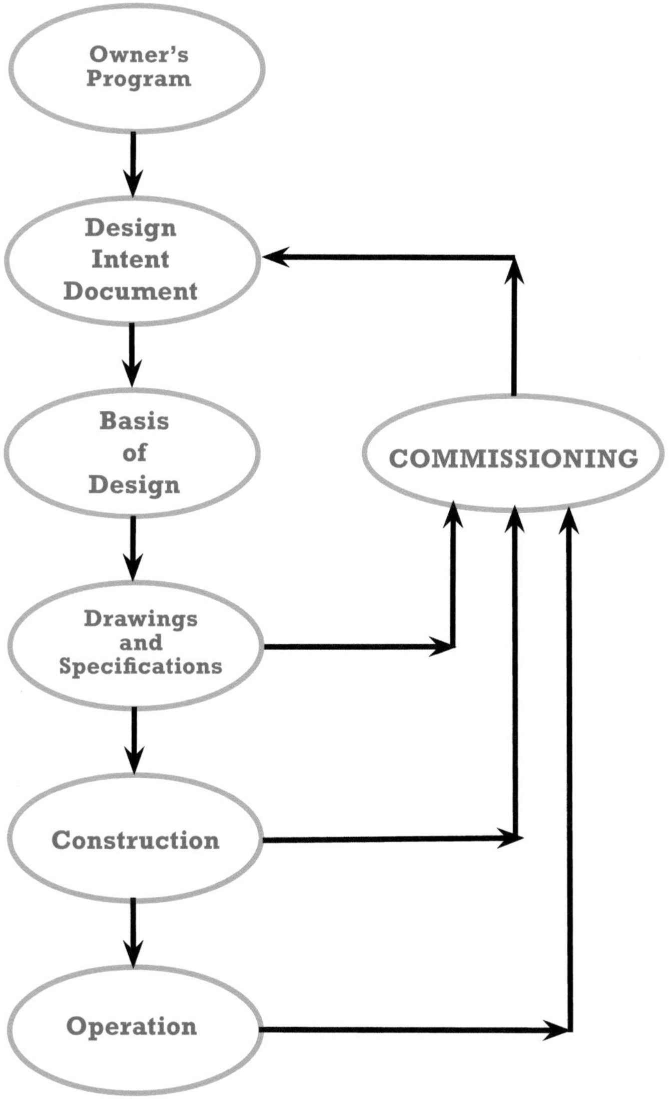

Figure 14.1

The DID sets the operational goals that form the basis for commissioning, whereas the construction documents set the requirements for the correct supply and installation of the building components. Together, the DID and the construction documents provide a complete description of component supply, installation and operation.

The document is architectural in origin and is developed, assembled and organized under the supervision of the architect. The document should be performance-based and concentrate on the occupants' needs rather than on how the design team will provide for those needs.

The DID should list the codes that are applicable in the area of the project. Since a Mixed-Use Building operation is intricate and voluminous in detail, the document should summarize the fire/life safety requirements of the design, and all other applicable codes should be cited and accompanied by a concise overview of the systems. Special systems should be included in the DID and in the scope of the commissioning process. Examples include security systems, as these systems are interlocked with HVAC and smoke detection systems.

Operator training needs to be addressed in the DID. Operating staff must be trained for this task and should be accomplished using the best possible methods.

14.5 The Commissioning Process

Commissioning takes place continuously during the building project, from programing through the first year of operation and touches all parties involved in design, construction, start-up, testing, operation and maintenance. The coordinator of the commissioning process is the commissioning authority (CA). The CA is hired by the owner and is independent of all other consultants and contractors on the project. This independence is important because the CA champions quality across all phases of the project.

Commissioning seeks to accommodate an owner's needs-needs that can change during construction or become apparent only after completion. This evolution often alters the original building system requirements; hence, the two primary functions of commissioning: testing and adjusting.

The process of commissioning requires that building components and systems be inspected and tested under actual installed conditions. As buildings become more complicated, the application of mechanical and electrical equipment to building systems becomes more convoluted.

Commissioning testing programs are different from those required for construction quality control, which prove that systems meet specifications based on original design conditions. Commissioning testing is employed when the building will operate under different conditions than anticipated during original design. These testing procedures demonstrate that the system is capable of meeting actual conditions or will point to adjustments required to accommodate the owner's use of the building.

14.6 Specifying Commissioning

The project architect should include the commissioning requirement as a separate project manual.

Requirements for coordination between the construction and commissioning efforts must be carefully drafted to avoid conflicts during construction contract closeout. The specifier should carefully define commissioning. This should incorporate other general requirements for commissioning, including:

- Utilities use during commissioning
- Products and systems changes by the commissioning agent, and the associated transfer of liability
- Voiding a manufacturer, subcontractor, or installer warranty
- Coordinating with the construction contractor during the correction period
- Reviewing maintenance manuals and making changes to the text
- Documenting changes made to equipment and systems

A separate commissioning contract also will require the owner to modify the construction contract's general conditions. The specifier should discuss the commissioning approach and related issues with the owner's legal counsel, who will also have to draft bidding and contract forms and conditions for the commissioning contract.

Commissioning services should be provided in five (5) phases: pre-design, design, construction, acceptance, and post-acceptance.

During the pre-design phase, the commissioning authority should carry out the following scope of work:

- Provide input to the Owner's requirements for the mechanical systems.
- Review the Design Intent Document and verify the initial design intent with the Owner and engineer.
- Prepare the pre-design commissioning outline.

During the design phase the commissioning authority shall carry out the following scope of work:

- Review the design documents (drawings and specifications) as they are prepared to ensure inclusion of material covering the contractor's responsibilities for commissioning; provide comments and suggestions for designer consideration.
- Prepare the design-phase commissioning plan.
- During the construction phase the commissioning authority should carry out the following scope of work:
- Organize and lead the commissioning team.
- Review shop drawings and equipment submittals for information affecting the commissioning process.
- Update the commissioning plan to reflect equipment and controls data from the submittals, and provide commissioning schedule information that the contractor can integrate into the project schedule.
- Schedule and lead commissioning meetings.
- Establish and maintain a system for tracking issues needing resolution.
- Review the project schedule periodically to ensure commissioning activities are properly incorporated; provide feedback to the designer as needed.
- Perform on-site observations during construction.
- Monitor correct component and equipment installation; including controls point-to-point checkouts. Document all observations.
- Witness equipment and system start-ups as deemed necessary. Ensure complete documentation of same.
- Other related work.

During the acceptance phase the commissioning authority shall carry out the following scope of work:

- Review and inspect, on a sample basis, the testing, adjusting and balancing work that has been carried out by another agency.
- Conduct functional performance testing of sub-systems, systems, and interactions between systems, leading to acceptance of the completed work. Document results of all tests witnessed.
- Organize and direct the training of O&M personnel.

During the post-acceptance phase the commissioning authority shall carry out the following scope of work:

- Conduct functional performance testing of sub-systems, systems, and interactions between systems that could not be carried out prior to acceptance due to unsuitable weather conditions.
- Prepare and submit a final commissioning report.
- Provide follow-up for quality performance during the guarantee period

The basic purpose of building commissioning is to provide documented confirmation that building systems are planned, designed, tested, operated and maintained in compliance with the owner's project requirements.

Direct and indirect benefits of commissioning buildings that factor into payback periods and returns on investment include:

- Savings in energy cost and improved building performance.
- Improved indoor air quality and comfort and increased productivity on the part of building users.
- Early detection of potential problems (the sooner problems are resolved, the less expensive they do to correct).
- Fewer changes during construction.
- Precise tune-up and operation of systems and applicable controls.
- Better building documentation.
- Trained building operators and maintenance personnel.
- Shortened occupancy periods.
- Reduced maintenance, operation, and equipment replacement costs.

14.7 Commissioning electrical systems

The overall goal of commissioning must be to ensure that a facility meets the design intent and the owner's requirements. This goal is generally achieved by proving to the owner that the reliability, redundancy, and resiliency that he or she paid for is indeed present and operational in the finished facility.

The commissioning authority has an obligation to provide a level of testing that will allow the owner to feel confident that each system is working and capable of maintaining a proper planned operational state during common external events.

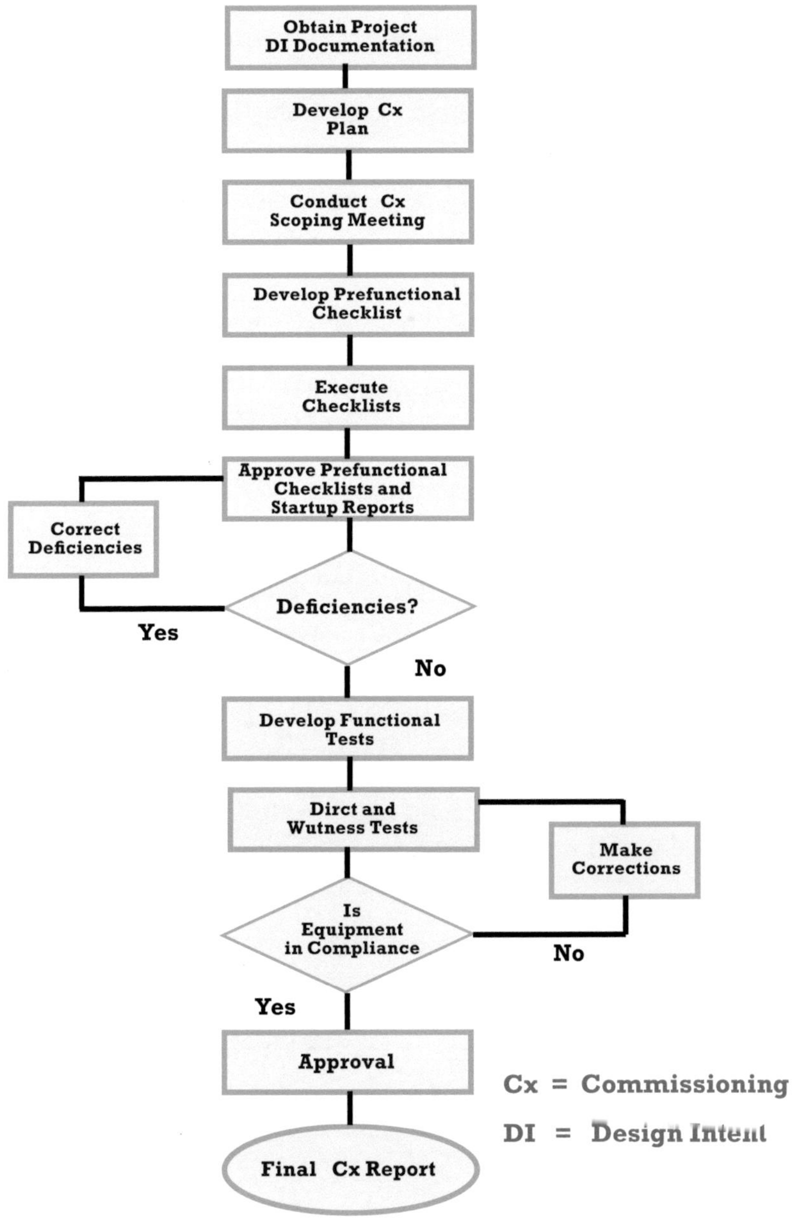

Typical Commissioning Process

A four-category rating system of over-voltage transient impulses (voltage spikes) has been established by ANSI, IEC (International Electrotechnical Commission) etc. and specifies general safety requirements for electrical equipment. In general, Category IV systems refer to power lines at the utility connection and the service entrance. This includes outside overhead and buried cable runs. Category Ill systems refer to distribution wiring, which includes 460 V 3-phase bus and feeder circuits, distribution panels, and permanent (hard-wired) loads. Category II systems refer to receptacle circuits and plug-in loads. Category I covers protected electronic circuits.

14.7.1 Thermographic surveys

Thermography is used for diagnosis and non-destructive testing of electrical equipment and power distribution systems. National Fire Protection Association (NFPA) 70B Recommended Practices for Electrical Equipment Maintenance provides guidance on setting up and performing an infrared scanning program.

Thermographic surveys should be performed during periods of maximum loading. Heating is generally related to the square of the current, so the load current has a major impact on ΔT. When conducting the survey, the engineer should record a description of the equipment, list the temperature difference between the component of concern and the reference area, and give the probable cause of the temperature difference. It is also essential to include load measurements for both the identified component as well as any reference component and include pictures and thermograms where possible.

ISO 18434 recognizes that thermal imaging is a useful tool to monitor the condition of equipment and wiring system intricacies such as unbalanced loads. Temperature abnormalities would indicate machinery overload, equipment misalignment or improper lubrication.

14.7.2 Harmonic Analysis

Non-linear loads in an electrical system can create power quality problems. In a Mixed-Use Building it is commonplace to have non-linear loads from equipment such as computer power supplies, solid-state lighting ballasts, variable frequency drives (VFDs), uninterruptible power supplies (UPS)etc. These non-linear loads generate voltage and current harmonics, which can have adverse effects on the equipment used to deliver electrical energy to the facility. Harmonics can have a detrimental effect on emergency or standby power generators and other sensitive electrical equipment. Harmonic analysis studies should be specified.

These harmonic-related losses reduce system efficiency, cause apparatus overheating, and increase power and air conditioning costs.

One of the most critical electrical systems to be commissioned in nonresidential buildings is essential/standby power equipment and system components. Given their importance, the NFPA provides specific requirements in NFPA 70: National Electrical Code (NEC), 2013; and NFPA 110: Standard for Emergency and Standby Power Systems, 2013.

With these requirements in mind, a standard commissioning process can be developed as outlined by ASHRAE Guideline 0-2013. The procedures, methods, and documentation requirements in this guideline describe each phase of the project delivery and the associated Commissioning Processes from pre-design through occupancy and operation, without regard to specific elements, assemblies, or systems.

Thermography can indicate the presence of harmonics in the form of overheated neutral conductors, bus bars, and transformers. A frequency other than 60 Hz (neutrals with harmonics will typically read 180 Hz) indicates that harmonics are present. Confirming harmonics can be done with multimeters but analyzing and mitigating them requires a power quality analyzer.

14.7.3 Lighting Commissioning

The current ASHRAE Standard 90.1: Energy Standard for Buildings Except Low-Rise Residential Buildings has significant lighting and control additions that require moderate ramifications from lighting designers and building owners.

ASHRAE 90.1 requires building lighting systems to be commissioned. A third-party commissioning authority, who is not involved in the design or construction, is required to verify that the lighting controls are adjusted, programmed, and functioning in accordance with the design and the manufacturer's installation instructions. The commissioning authority must also submit documentation certifying that the lighting systems are in compliance with or exceed the performance requirements. The certification needs to be specific enough to verify conformance to the authority having jurisdiction. Requiring the systems to be commissioned is the best way to ensure that they operate as intended and that the energy-savings strategies have the best chance of realizing actual reduced energy consumption.

14.7.4 Whole-Building Shutdown Tests

To ensure emergency systems in Mixed-Use Buildings will function properly when called upon, it is essential to conduct an intentional power outage and recovery. In order to test the systems, power to the entire building need to be shutdown and restarted. Every piece of equipment that is electrically powered is involved in a whole building shutdown test, whether it is on standby power or not.

Experience has shown that the designer, contractor, and facility management staffs have very different understanding of what should actually happen during a power outage. These expectation differences occur primarily because the construction documents are not clear. Confirming these expectations through a formal test is critical to ensure that life safety, product and equipment protection, comfort are not compromised during a power outage.

During the whole building shutdown test the electrical power to the entire building is should be shut off to imitate a "utility power" outage. The standby electrical generator is allowed to start normally, and later normal utility power is restored. During the loss and restoration of power equipment reaction must be observed and documented against the specifications and an event matrix. An event matrix is a table that lists each piece of equipment and how it responds to power loss and restoration. The matrix should be provided by the designer and commissioning authority.

Because the response to power loss involves so many disciplines, coordination is essential. It also requires a champion. The commissioning authority who has a more global perspective of all building systems should manage cross-discipline issues to resolution.

14.8 Commissioning HVAC Systems

Today's HVAC systems must be energy efficient, satisfy

stringent indoor air quality and comfort expectations, and still be designed and constructed within tight budgets. System designs meeting these demands typically have many components, sub-systems, and controls.

HVAC systems installed under contract are to be inspected, tested, signed off as complete and operational, and operated for commissioning authority verification. The foregoing should include a list of components, equipment, and systems that must be commissioned. The following list is generic and should be edited and extended as appropriate for the specific project:

- Hot water, chilled water, and condenser water piping system
- Duct and air-handling systems
- Chiller(s)
- Cooling Tower(s)
- Refrigeration Compressor/Condensing Unit(s)
- Boiler(s)
- Pumps
- Supply, Return, Relief and Exhaust Fans
- Air Handling Units (both packaged and built-up)
- Air Terminal Devices
- Fan-coil Units
- Water-source Heat Pumps
- Direct digital controls system
- Building Management System

14.9 Commissioning Non-HVAC Systems

There are many non-HVAC systems which benefit from commissioning. Some of these systems are related to HVAC in that they include mechanical or ventilation components or controls interactions, but their function is primarily oriented to process requirements or safety, not occupant comfort. Other systems are in completely different disciplines.

Systems that are life safety related will be governed by regulations or codes that mandate specific, formal, test procedures. Among the systems that are frequently commissioned are the following:

- Fire protection
- Fire alarm
- Smoke control
- Security
- Elevator control
- Space pressurization

14.10 Project Closeout

Just as it necessary for the owner to have an operating and maintenance manual, it is necessary to have a report of all the changes that were made to the project, and the reason they were made. Included in this report should be computer simulations and analyses to document the changes.

Since changes during construction are also a fact of life, the design plan may have to be modified to match the final construction, and these "As-Built" plans reviewed to ensure their accuracy.

The operations and maintenance manuals should include a complete detailed maintenance program for the facility as constructed, including methods and frequency of service, lubricant types, inspection procedures, and maintenance log forms. A special section of the manual should include manufacturer's detailed equipment sheets, a list of spare parts that should be inventoried by the owner, the names and addresses of the equipment manufacturers, and service agencies.

14.11 Post Construction Cleaning

Every Owner expects his/her new property to be spotless and perfect on move-in day. Tight construction and occupancy deadlines make these expectations difficult and challenging to meet.

A professional and experienced Post-Construction cleaning services provider should be appointed. In general, the main contractor should plan ahead with a professional cleaning services provider to perform the post-construction cleaning.

Construction follows a specific workflow. Flooring is installed early in the buildout process and should be adequately protected as the project progresses. For example, newly installed flooring can take a beating as contractors and trades workers continuously walk through and work in the space. Dust and debris end up on all surfaces with paint and adhesives as well as chemicals and other and materials.

The cleaning services company should effectively remove construction dust from surfaces, polish glass, and clean up all of the debris. Specialized equipment's are needed to clean both textiles and hard surfaces – including walls – properly. Additionally, highly trained technicians who know the appropriate chemistries, equipment and methods to clean according to product specifications should be employed. They do more than clean – they should be able to repair damage, improve the appearance, and safeguard against future mistakes.

The advanced cleaning equipment and cleaning, sanitizing, and disinfecting solutions should include the use of third-party (Green Seal, EcoLogo, Design for the Environment) certified products. Just as the correct cleaning equipment is essential, so are the right chemistries and cleaning products. Specialty building materials require unique care. Specialist can reduce long term maintenance costs by recommending performance coatings or other treatments for post installation application.

Significant cost savings can be realized if the cleaning process is proactively planned to align with project costs and timelines. It is essential that post-construction cleaning is allocated in the construction budget.

Section 15.0

Facility Operations

15.1 A Proactive approach for long-term success

The success of a sustainable building relies on sustainable design, green building construction and sustainable operations. Sustainable operations continues through occupancy by a proactive approach.

The moral, ethical, social and political arguments for taking action on environmental issues are becoming ever more persuasive - and more widely accepted. But actions by businesses to improve environmental performance are unlikely to be motivated by altruistic reasons alone. Both management and stakeholders need to hear clear commercial reasons for investing time and money in environmental initiatives. The environment now interests people in all walks of life throughout the world.

The environment presents a variety of new opportunities for businesses, including opportunities to save on the cost of day-to-day overheads. Benefits of taking action include:

- Lower utility and purchasing costs - by using what you buy more efficiently.
- Lower disposal costs - by avoiding excess packaging and sorting waste for recycling.
- An opportunity to attract the new breed of 'green' consumers.
- A safer workplace for your employees, reducing the risk of health and safety problems.
- Boosting staff morale and reinforcing a commitment to quality.
- Improving relations with the local community.

The commercial penalties for hotels that fail to act on the environmental agenda include:

- Loss of market share from a poor public profile.
- Risk of fines from failing to meet legislation.
- More difficulty attracting and keeping shareholders.
- Risk of tarnishing your brand image.
- Higher turnover of staff arising from failure to engender pride in the company.
- Not being prepared for steep increases in costs.

In addition to a localized impact on the environment, facility operations contribute to a variety of global environmental problems. Some of these issues are:

- Global warming arising from an increase in greenhouse gases, such as carbon dioxide, in the atmosphere.
- Ozone depletion in the upper atmosphere from a build-up of chlorofluorocarbons (CFCs).
- Acid rain, whose causes include the burning of fossil fuels to generate electricity.
- Exhausting the earth's remaining supplies of fossil fuels and mineral deposits.

While one facility's contribution to alleviating these problems may appear relatively small, when large numbers of facility operations all act together the aggregate benefits are substantial. Such improvements contribute to the long-term sustainable development of the industry - essential in view of the growth in demand predicted for the next few decades.

Facility operations which are taking steps to monitor and reduce energy and water consumption, to use resources efficiently and to reduce waste, find that these measures pay for themselves through cost savings - and improve the competitiveness of the business.

Success in business depends on developing and maintaining a strong market position. Concern for the environment can help to secure the goodwill of all stakeholders: employees, suppliers, guests and the wider community. Especially in markets where a hotel group is not yet well established, the association of the company's name and logo with projects aimed at benefiting the environment can boost brand image and competitive market position.

There is a growing trend towards stricter environmental legislation all around the world. Taking action now, in advance of greater regulation of hotels, will make it easier to meet new laws when they do come into force. It may also mean that future regulators are less demanding of the hospitality industry; because they already see evidence of good environmental performance.

15.2 Environment as a business issue

The environment as a business issue is little different to other business issues, requiring the same strategic approach from management, including understanding the effect it has on the business, developing a policy for dealing with it, and setting in place a process for implementing the policy.

The complete process is shown Figure 15.1. It shows elements of key points at a glance and indicates the potential way forward.

Figure 15.1

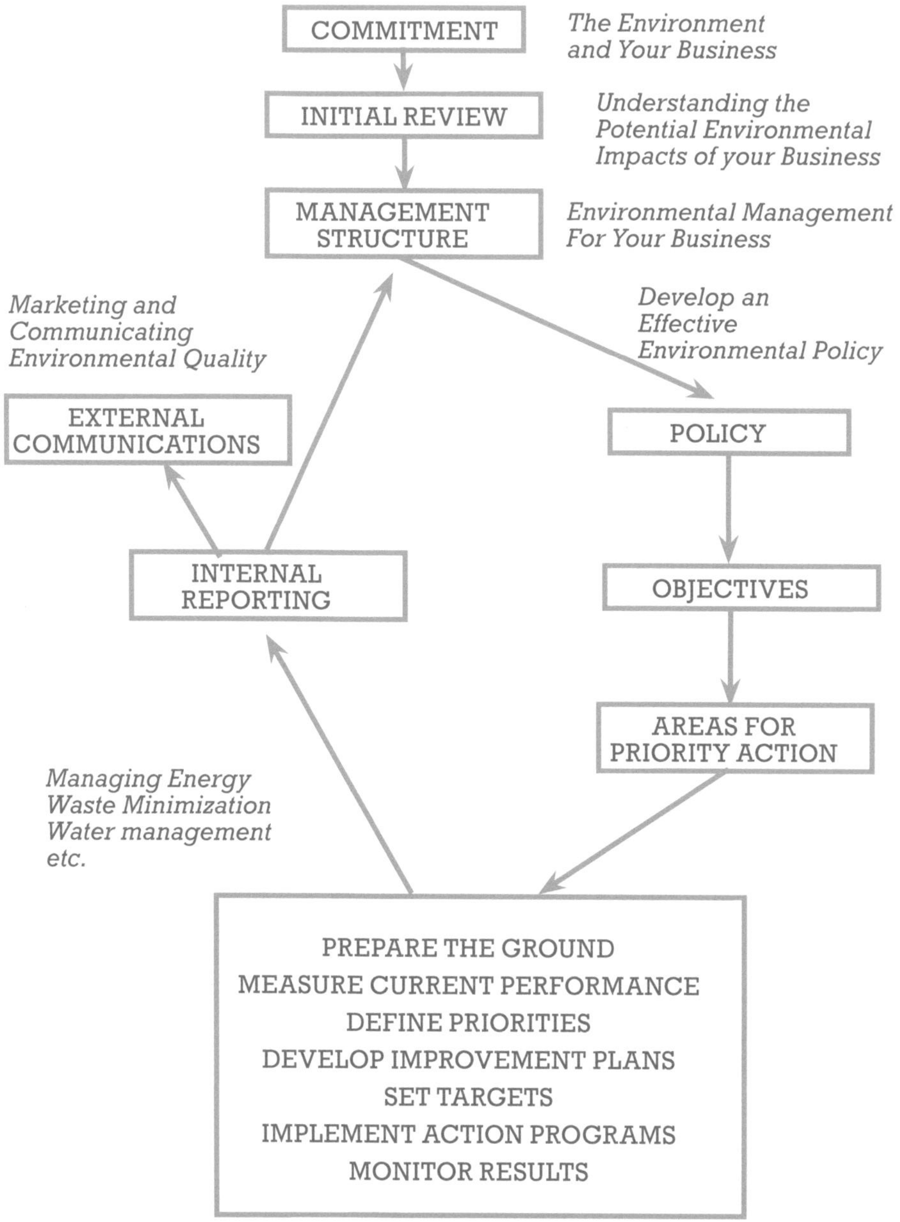

The starting point of an environmental program is to identify environmental impacts of one's business activities and address them in a positive way. The following are intended to identify potential requirements:

15.2.1 Organization Structure

- Company Policy

  Consider if the availability of the structure to implement the environmental policy. Develop a written policy for the environment, agreed at the highest level of stakeholders.

- Responsibility

  Appoint a senior company official with direct responsibility for environmental issues. An environmental committee should be appointed to report to the designated individual, with representatives from all areas of the business to oversee the implementation of agreed plans.

- Communication

  Communicate the environmental policy internally and externally. The communications should reach different groups of stakeholders in different ways, such as:

  * Customers via regular bulletins at points of sale, in brochures, etc. Promoting local sustainable shops, activities and alternative means of transport encourages the customer's environmentally conscious choices to permeate into the surrounding economy.
  * Shareholders, bankers and insurers, via specific references to the environmental program in the company annual report.
  * Local communities via local press.
  * Employees via in-house magazines, notice boards, consultation meetings.
  * Media, via press releases and more formal press briefings.
  * Environmental groups, via specific information packs.
  * Local regulatory officials
  * Staff Training

Many of the sustainable measures put in place by the design and construction teams are only effective if they are properly implemented by the staff and building operators. Emphasis should be placed on providing knowledge to the environmental committee head and to the facility management's training manager. The primary goal of the training would be to create awareness of the environmental issues, and secondly to initiate and follow through appropriate programs.

It may be necessary to engage the services of an environmental consultant for development of action programs.

15.2.2 Operations

A transparent and sustained management commitment can help secure employee commitment towards sustainable operations. The subject of environmental knowledge is extensive in its scope. As one's knowledge grows, the individuals would gradually understand the enormity of the consequences of human beings' actions that damage the environment.

Environmental responsibility is manifest for each person, organization and society. This shared ecological responsibility has caused governmental entities to establish laws stating standards of conduct in operations. In all cases, the relevant local, national and regional regulations should be followed, as must the relevant corporate standards. In cases where several alternative regulations or standards apply, the most stringent regulation should be taken as the minimum acceptable standard.

Where the services of contractors/haulers etc. are engaged, the contractor/hauler should indemnify all related building owners, and they should also provide evidence of sufficient and comprehensive limits of liability insurance against pollution and/or exposure to toxic substances.

The overall objective should be to maximize the efficiency of the facility operation while minimizing the overall environmental impact.

a) Preparing the ground

Establishing a successful environmental program in the building facility will require actions on a number of fronts. The main thrust of the program will be to influence and improve day-to-day operations behind the scenes. The management systems approach should be based on clear policies, reviews of activities, the setting of targets and periodic audits of achievements. By planning actions, motivating employees and monitoring progress would facilitate achieving your objectives.

Keep in mind that mixed-use buildings are at the center of a huge network of people - staff, customers, business partners and the local community - all of whom influence and are influenced by facility management actions. Many of the processes that need to be implemented will be impossible without their goodwill and co-operation. Communication of the environmental program to all these people will increase its effectiveness, particularly as they take the message into their own organizations. This creates an enormous opportunity to raise awareness and influence a vast number of people, while at the same time demonstrating to the outside world that the mixed-use facility recognizes the importance of the environment and is taking the necessary actions to protect it.

Good preparation is essential to the successful outcome of the review program. In order to implement the program, facility management would need to set up an environmental working group, whose job it will be to inform, support and motivate the entire staff. How you do this will depend very much on the size of the property and what staff are available. To make a success of this program a figurehead or 'champion' should be appointed to drive home the message -someone to ensure that all stakeholders understand how the policy affects them.

Most of the review operations could be conducted internally, using staff with sufficient knowledge of the business. However, there may also be a need for detailed technical assessments of some issues, requiring the assistance of external consultants. Alternatively, time constraints or the need to have an impartial overview may necessitate the use of external consultants for all or part of the operation.

Systems for the environment may be easier to establish when other management systems are already in place. There are various different models for establishing such systems, but they typically cycle through each of the following stages:

- Policy
- Reviews
- Objectives

b) Policy

The environmental program requires the setting out clearly and concisely - program aims. An environmental policy, or mission statement, outlines these aims and the principles to follow. A formal policy of this type is invaluable in telling your stakeholders - employees, customers, business partners, local residents - that facility management is committed to environmental improvement.

The policy should:

- Demonstrate commitment to good environmental practice.
- State facility management aims and provide a summary of how you plan to meet them.

It should also be supported by an action plan:

- Specifying short-term targets and medium- to long-term objectives.
- Showing how and when these will be achieved.
- Naming staff members responsible for the actions necessary to meet the objectives.

The policy should be revised and updated periodically to take account of progress and new priorities.

An environmental policy is one of the key elements of an environmental management system, and a requirement of the international standards for environmental management:

ISO 14001-2015 and its supporting standards such as ISO 14006:2011 the European Eco-Management and Audit Scheme (EMAS).

The International Standard ISO 14000 series complements the ISO 9000 series - the quality management standards – in as much as they take parallel approaches to achieving and demonstrating compliance. Organizations with a management system that meets the requirements of ISO 9000, should be able to extend it to cover environmental objectives and so qualify for ISO 14000.

Both EMAS and ISO 14000 are voluntary schemes and do not include specific environmental performance criteria. Instead, they direct participants to set their own objectives and commit themselves to continual improvement. Documentation is essential for organizations that wish to be certified under the schemes.

c) Environmental Review

A review is usually one of the first steps towards establishing an environmental management system. It should address the main environmental impacts of the building facility and opportunities for reducing them:

- Waste management
- Product purchase
- Indoor air quality
- Air emissions
- Energy conservation
- Noise
- Storage tanks

- PCB's
- Pesticides and herbicides; Hazardous materials
- Water
- Community action

Many of the sections inter-relate and overlap. In the case of "Product purchase" for example, all sections are relevant, and will have a bearing on the eventual building operation purchasing policy.

d) Auditing

International environmental standards make a distinction between audits and reviews. A review is undertaken early on to flag up areas for attention, whereas an audit happens at the end of each cycle through the program, typically at the end of each year. The purpose of the audit is to check whether targets and objectives have been met and to see what improvements have been made. Measuring environmental improvement like this helps to maintain enthusiasm for an environmental program -especially when the organization meets its objectives. It is essential to communicate the results of such a check to all stakeholders.

e) Environmental working group

In order to implement the program, facility management will need to set up an environmental working group, whose job it will be to inform, support and motivate the entire staff. How you do this will depend very much on the size of the property and what staff are available.

The program will only succeed if a figurehead or 'champion' is appointed to drive home the message - someone to ensure that all employees understand how the policy affects them.

This coordinator will:

- Implement the environmental policy.
- Ensure that objectives are set.
- Keep the program moving forward.
- Encourage everyone to get involved.
- Supervise the auditing of targets.
- Collate information, ideas and results and communicate them throughout the facility.
- Communicate results of the environmental program externally, to local authorities, local residents, other interested parties and suppliers.

He or she will chair the meetings of the environmental working group or committee. He or she should have a good operational knowledge of the business, the respect of other staff, a commitment to the project and full support from the facility manager.

f) Business partners

The range of goods used by the mixed-use building facility is vast and covers most of the manufacturing base of an economy, from building materials to furniture and fittings. These facilities also use the full range of professional services, from advertising agents to merchant banks, and communicate via diverse media. This interdependence with so many other businesses gives the facility manager an unique opportunity to challenge business partners to help in the environmental mission and to look at how they might change their own activities to be more environmentally conscious.

Acting together, the facility operator could be a formidable force in influencing a substantial and diverse body of people. The more facility operators demand environmentally friendly goods, the greater the market will be and the cheaper the goods will become to obtain, which would be good news for everyone as well as for the environment.

There are solid business reasons for adopting an environmental approach, not least the fact that environmental considerations are becoming increasingly important for building occupants. It is vital that the customers and residents are made aware of schemes and the reasons for those initiatives. Good public relations are an indirect way of recouping some of the investment needed to make improvements to some aspects of facility operation and is an essential part of spreading the word and encouraging similar action by competitors.

g) Implementing Action Programs

Main findings of the review should be addressed as an executive summary incorporating outline plans for implementing the relevant processes based on appropriate targets. Drawing on the information collected, priorities should be definod by the review team for each area selected for action.

15.3 Action Programs

15.3.1 Waste Management

From a sustainability perspective, efficient waste management practices of the mixed-use facility is a pivotal part of its overall green strategy. Waste disposal creates a myriad of environmental problems that may not be reversed for many generations. Carbon emissions from transportation, habitat depletion for landfills, airborne pollutants from trash incineration, toxins leaching into water supplies and greenhouse gas emissions from decomposition are some of the environmental impacts of waste disposal. With millions of tons of construction debris buried in landfills every year, waste reduction programs offer many opportunities for carbon reduction and habitat protection.

Waste reduction and diversion programs begin at construction and continue through to operations. Comprehensive programs should encompass the supply chain waste reduction and recycling/reuse programs. Understanding a facility's waste stream contributions will present many opportunities for reductions in landfilled trash as well as disposal fees.

The first step in towards the effective management of waste is to identify the types and quantities of waste being generated. The facilities management team should conduct a waste audit. A waste stream audit is a process used to quantify the amount and types of waste being generated during ongoing operations, as well as the treatment and disposal method of the wastes. identifying recyclables in the waste stream that would go to landfill can lead to reduced waste hauling costs. Weight and volume play an important role regarding waste audits.

Tho head of the Environmental Committee should inquire about local waste management resources to determine what recycling services are available and the associated costs. Successful implementation of recycling programs begins with the education of all stakeholders. Customers and building occupants should be made aware of how to participate in the measures that have been established by the facility. Staff should be trained on separation and storage policies for recyclables.

Facilities must also be in place to support this program. At a minimum, the building facility should have a central storage area to allow for waste separation. Recycling bins throughout the property not only facilitate waste separation but also make these efforts visible to visitors and building occupants.

Waste from a mixed-use facility could consists of both wet (organic/biodegradable) and dry waste. The wet waste consists primarily of food waste, which can account for substantial percentage of total waste as collectively they account for a very significant proportion of waste generated. This would depend on the extent to which the establishment operates restaurants, food halls, cafeterias etc. in addition to residential apartments.

Food waste could be any food discarded as part of operations in the hospitality and residential sector. This could be packaged food that is thrown away as it has passed its expiration date, food preparation/processing wastes, and waste from serving dishes and customers' plates.

The components of dry waste could be classified as non-hazardous and hazardous types of waste. Figure 15.2 and 15.3 describe the components of non-hazardous and hazardous types of waste, respectively. It should be noted that these are not exhaustive lists of the components, although they mention all of the most significant components. For instance, sometimes mixed-use buildings produce other types of wastes, such as bulky items (e.g. furniture), construction and demolition waste (e.g. concrete, pipes, etc.), discarded electronics and office appliances, and used refrigerating equipment. The relative percentage of each of these different waste types could vary by establishment.

Fig. 15.2 Non - Hazardous Waste

Waste Type	Components
Household Waste	Food/kitchen waste, used or dirty paper and wrapping, plastic wrapping or bags, composite wrappers
Cardboard	Packaging
Paper	Printed documents, brochures, menus, maps, magazines, newspaper
Plastic	Bags, bottles (that did not contain hazardous material), household goods, individual portion wrappers for various products
Metal	Tin cans, jar lids, soda cans, food containers, mayonnaise, mustard and tomato puree tubes, aluminum packaging
Glass	Bottles, jars, flasks
Cloth	Tablecloths, bed-linen, napkins, clothes, rags
Wood	Wooden packaging, pallets
Organic waste	Fruit and vegetable peelings, flowers and plants, branches, leaves, grass

Fig. 15.3 Hazardous Waste

Waste Type	Source
Cooking oil	Kitchens, restaurants
Mineral oil	Maintenance service
Paint and solvent residues	Maintenance service
Flammable material (gas, petrol, etc.)	Kitchen, garden, Maintenance service
Fertilizers and chemicals (insecticides, fungicides, herbicides)	Landscaping
Cleaning chemicals	Maintenance service
Ink cartridges	Administration
Batteries	Maintenance service, administration, residential apartments
Cleaning chemicals and solvents	Housekeeping and laundry operations
Electric bulbs	Maintenance service

15.3.2 Waste Audits

Waste auditing is a strategy which is used by organizations to facilitate more effective waste management. It helps the establishments understand where and how waste occurs. waste auditing involves monitoring waste generation at a property in terms of which types of waste are generated, in what amounts, and in which locations. A waste map is subsequently created to reflect this data. Facility Management can subsequently plan its waste management operation in a more efficient manner.

15.3.3 Product Purchase

Environmentally responsible or 'green' procurement is the selection of products and services that minimize environmental impacts. Green product can be categorized as a product that will not pollute the earth or deplore natural resources and can be recycled or conserved. Some examples of these products are "household items manufactured with post-consumer plastics or paper, recyclable or reusable packaging, energy-efficient light bulbs and detergent containing ingredients that are biodegradable, non-polluting etc. Common terms normally used by companies promoting green products are "eco-friendly', "environmentally safe", "recyclable", "biodegradable" and "ozone friendly".

Purchasing decisions can make a significant contribution to environmental protection. The concept of 'precycling' - choosing products that are environmentally friendly in terms of content, packaging, ease of recycling and disposal - informs good purchasing practice. In purchasing responsibly, an organization becomes more cost-effective, and by stimulating the local economy enhances the local environment. On the global scale such purchasing by organizations would impact beneficially on the supply industries.

Principles of responsible purchasing

a) **Set your goals**

- Purchasing sustainable products would ensure that the company is meeting national, international and own sustainability goals. Therefore, ensure that procurement is a key part of the overall sustainability plan.
- Install a sustainable procurement plan with measurable goals that cover all major product categories.
- To decide which product categories are most relevant to include in the sustainable procurement plan, evaluate environmental and social risks during the whole life cycle, and also potential gains that purchasing more sustainable alternatives can lead to.
- Buy only what you really need.

  Reducing consumption is the first step. Scrutinize all purchase requests and only purchase what is absolutely necessary.
- Buy quality.

  Buy the highest quality items affordable and have them repaired and serviced when necessary. High quality durable goods don't need to be replaced as often as low-quality goods; they save money and resources in the long run.
- Buy locally.

  Transportation costs should be a factor in purchasing decisions. For example, if good quality food products are available in the region, why purchase from a far-off location. Also, locally purchased food items tend to be less processed and use fewer preservatives.
- Buy in bulk.

  When buying products to be consumed in quantity, buy the largest amount that can be economically stored and used. Not only does bulk buying reduce the amount of packaging produced, it also saves energy by reducing deliveries resulting in overall cost savings. Clearly the precise amount to be purchased needs also to be assessed based on economic factors.
- Buy for energy efficiency.

  Use of energy labels in the procurement processes. Energy efficient office equipment, appliances, lighting products, technology products, household appliances, etc.
- Buy recycled or recyclable products.

  Where possible, buy products that are made from recycled material or are recyclable. The most common recycled goods are in the categories of paper, glass and aluminum.

b) **Involve the right people**

- Ensure that management and co-workers are aware of the social and environmental risks connected to products and how sustainable procurement enables the company reach organizational, national and global sustainability goals.
- To be able to work strategically and long-term with sustainable procurement, ensure management-

level approval is in place on both the sustainability plan and the sustainable procurement plan.

- Include team members with sustainability expertise in the procurement process to ensure that these aspects are considered before, during and after contracting.

c) **Clarify your intentions**

- Explain to all vendors that company social and environmental sustainability goals are a priority aspect of the purchasing program.
- It must be clearly stated that all technology products are a prioritized product group in sustainable procurement.
- Avoid disposable products

  Disposable, or "throw-away," items are generally not a good environmental option. This is because of high level of energy and resources that go into manufacturing and transporting these one-time use products. Even if disposables could be recycled or composted, the majority of their environmental impact occurs "upstream" – in manufacturing and transportation - before they are even used.
- Minimize packaging

  While packaging performs the key functions of protecting a product and providing information for buyers, what we tend not to think about is what happens to it later. When it comes to disposal, most packaging is not recycled, but dumped in landfills or burnt. This results in pollution in a number of ways.

  The best way to reduce packaging materials going to landfills continues to be through the use of lighter weight packaging. Recycling of primary packaging plays a prominent and growing role in reducing discards.
- Source reduction continues to play a significant role in the effort to reduce material usage and waste and therefore need to take center stage because recycling could apparently not grow fast or large enough to offset increases in waste generation.

d) **Involve vendors early**

- Generally, brands apply for product certification when purchasing organizations requests for approved certification. Make it known to vendors early that the company intend to specify the latest generation of certification in the purchase contract terms.
- Ensure that the contract name reflects the requirements that are most important to you, as well as the product category that the contract will cover, for example: "Procurement of sustainable SMART Boards". Naming it correctly will enable a smooth procurement process, where you are more likely to receive the right offers from vendors
- Favor products with an independently awarded eco-label. Include the latest generation of certification for the product category in the contract document and require a valid certificate as proof of compliance..
- Look for unbiased evaluations and certification of products to make informed, responsible green choices. The most common and respected certifying

bodies currently working in North America are:

* **Cradle to Grave Certification -** Provides certification for a variety of building and consumer products
* **EcoLogo Program –** This organization certifies large number of products, including appliances, gardening supplies, home improvement products, and cleaning supplies.
* **Energy Star -** They certify a variety of products according to the energy they use, including home entertainment, computers, laundry and kitchen appliances, and more.
* **Green Label and Green Label Plus** - Administered by The Carpet and Rug Institute, these certifications identify products with very low VOC emissions
* **Green Seal** - An independent, non-profit organization that provides science based environmental certification standards for anything from cleaning services to paints.
* **GreenGuard** - Certifiers of many green products, including furniture, paint, flooring and insulation.
* **Scientific Certification Systems (SCS)** - An independent, third-party certifier. They administer certifying systems on a variety of products.
* **SmartWood** - The largest **Forest Stewardship Council** (**FSC**) certifier of forestlands. Look for their seal of approval on wood, paper, and forest products.

- Include a qualification period (for example three months) in the contract, to allow vendors to obtain certification
- Specify that products must maintain an active certificate throughout the entire contract period — even if the manufacturer refreshes or updates that product during the life of the contract.
- Include a contract severance clause or other consequences in case a vendor does not provide valid certificates as proof of compliance during the life of the contract.

Consider which products, services or works are the most suitable on the basis both of their environmental impact and of other factors, such as the information available, what is on the market, the technologies available, costs and visibility.

Draw up clear and precise technical specifications, using environmental factors where possible

Establish selection criteria: Where appropriate include environmental criteria to prove technical capacity to perform the contract. Advise potential suppliers, service providers or contractors that they can use environmental management schemes and declarations to prove compliance with the criteria.

Establish award criteria: where the criteria of the 'economically most advantageous tender' is chosen, insert relevant environmental criteria either as a benchmark to compare green offers with each other (in the case where the technical specifications define the contract as being green)

or as a way of introducing an environmental element (in the case where the technical specifications define the contract in a 'neutral' way) and giving it a certain weighting. Consider the life-cycle costing.

Use contract performance clauses as a way of setting relevant extra environmental conditions in addition to the green contract. Where possible, insist on environment-friendly transport methods.

e) Sustainable Terms and Conditions with Business Partners

Sustainable business practices are shaped by the relationships formed with suppliers and partners. Business partners should be chosen based on the degree to which they support the social and environmental goals adopted by the company. Facility Management should look for opportunities to work closely with product and service providers.

The criteria used in selecting vendors and suppliers should include environmental considerations, pricing, reliability, reputation and service. Clear standards of delivery and disposal of waste should be established with main suppliers in order to reduce amounts and oblige suppliers to work out a solution for their own disposal of waste.

Developing a questionnaire for prospective suppliers will assist in evaluating which suppliers meet your organizations environmental standards. In particular, buyers should aim to find out from suppliers:

- Whether products contain recycled materials.
- How products will be packaged.
- Whether products have toxic effects.
- What potential there is for recycling.
- Whether products are produced locally.
- How durable the products are?
- How the price of environment friendly products compares with alternatives.

A detailed picture of the environmental impact of purchases should be developed by asking questions relating to each stage of the product's life cycle - from 'cradle' to 'grave'. That is, to find out what environmental impacts arise from extracting the raw materials, manufacturing the product, transporting it, using it and disposing of it.

f) Life cycle costing

As sustainability is becoming a powerful growth engine consumer goods manufacturer are considering their total environmental impact—from growing raw materials to the consumer's disposal of their product.

- The initial purchase/commission cost.
- Running costs.
- Maintenance costs.
- Ultimate disposal costs.

This approach gives a more accurate assessment for comparing the costs of different products. It also means that products with a lower environmental impact those that consume less energy or water or are less dangerous to dispose of - are not rejected simply because the initial price is higher.

Fig 15.4 Purchasing life cycle assessment checklist

Stage of lifecycle	Environmental issue
Raw materials	Originate from environmentally sensitive areas?
	Damage local environment when extracted?
	High energy input necessary for extraction?
	Long distances and inefficient modes of transport for materials?
	Low recycled, post-consumer waste content?
Manufacture	Exploitative employment practices during extraction?
	High energy input?
	Large volume of waste generated?
	Large volume of effluents generated?
	Toxic air emissions released?
	Releases heavy metals?
	Uses solvents?
Transport and delivery	Long distance travelled from manufacturer to facility?
	Inefficient modes of transport over this distance?
	Possibility of hazard during transport?
	Excessive use of packaging?
	Supplier does not collect used packaging to re-use/ recycle?
In Use	High energy consumption?
	High water consumption?
	Air emissions or toxic effects?
	Short length of service before need for replacement?
	Likely need for - and/or problems obtaining - replacement parts?
Disposal	Presents some threat to the environment at end of useful life?
	Special requirements for safe disposal?
	Little potential for re-using all/most/ some material?
	Little potential for recycling all/most/ some material?
	Impossible to return product to supplier for re-use?

15.3.4 Indoor Air Quality

The use of mechanical ventilation has tended to increase the residence time of internal air and this section is intended to provide an interior environment that is conducive to the health and well-being of building occupants.

a) Potential sources of air pollutants

Special consideration should be taken since contamination levels in a building are directly influenced by the type of activities that take place within the building. For instance, IAQ in office buildings is affected by the emissions from office furniture, office materials and equipment such as copiers and printers. Similarly, IAQ in food establishments are affected by the humidity, fumes, odor generated by cooking, and environmental tobacco smoke (ETS) from smoking room infiltrating into other parts of the building.

- Combustion products

  These may include gases (such as carbon monoxide, nitrogen oxides, Sulphur dioxide or hydrocarbons) and suspended particulates from boilers, cooking stoves, vehicle engines in garages and other combustion sources.

- Chemical vapors

  These may come from cleaning solvents, pesticides, paints and varnishes.

- Building materials

  Such materials may include toxic substances, such as formaldehyde in foam insulation, textile finishes or pressed wood, fiberglass or mineral fibers, plasticizers, etc.

- Airborne micro-organisms

  Such organisms as legionella are primarily associated with moisture in the air conditioning and ventilation systems. Droplet infection is an issue in inadequately ventilated and crowded places.

- Dust or particulate matter

  Introduced with outside air or from internal activities, they may also contain micro-organisms, and can be irritants, particularly to people with allergies or respiratory weaknesses. They can damage equipment and decor and will increase cleaning requirements.

- Radon gas and radon products

  These are released by the soil on which the building is situated or by stone, especially granite, cement or brick building materials. Radon most commonly enters the building in the form of soil gas that is drawn in through joints, cracks or penetrations when the building is at negative pressure relative to the ground.

- Methane gas.

  This may come from decomposition of landfill material if the building site is on or near a landfill for municipal waste.

Good IAQ depends on addressing successfully questions of ventilation, humidity, odors and chemical vapors.

b) IAQ management program

An active IAQ management program specific to the building should be developed to achieve long term IAQ goals.

- Develop an IAQ profile of the building by review all available documents and/or records related to the design, construction, operation and maintenance of the building and the air-conditioning and mechanical ventilation (ACMV) system.
- Develop and implement plans for the operation, preventive maintenance and unscheduled

maintenance of the ACMV system and housekeeping activities. The IAQ program should include routine review and correction of problems in the areas listed below:

- Cooling towers -Eliminate stagnant water accumulation and remove biological contaminants.
- HVAC supply ducts and cooling coils - Check for cleanliness, eliminate excess moisture, remove micro-organisms and particulate matter.
- Ventilation rates (particularly in variable air volume systems) - Maintain required outside air supply and distribution. Ensure that all systems are properly balanced and a surplus of approximately 10% is maintained, to prevent negative pressure within the ventilated space, with the exception of a kitchen and laundry. Control possible pollution source via fresh air intakes from the outside, i.e. traffic emissions, boiler flues, cooling towers etc.
- Use of pesticides - Make proper selection, control storage and application and ensure adequate ventilation
- Use of cleaners and solvents - Follow manufacturers' guidelines and ensure adequate ventilation
- Routine cooking facilities operations - Check control of fumes and odors and ensure proper ventilation
- Heating, boiler or other combustion systems - Check proper air-fuel ratio, stop fuel leaks and ensure proper venting of waste gases
- Loading bays - Prevent vehicle emissions from being drawn into the building. The same applies to all odors from the waste removal site
- Storage facilities - Follow manufacturers' instructions on proper storage of items, especially those that release odors or irritants
- Battery rooms - Control the formation of explosive and acidic gases
- Parking areas - Control ventilation rates and prevent vehicle emissions from entering the premises

- Develop and implement procedures for dealing with building renovation, addition and alteration, pest control and other activities that may have an impact on IAQ.
- Inform building occupants about their activities that may impact IAQ and what they can do to maintain acceptable IAQ.
- Establish clear procedures for recording and responding to IAQ complaints and inform FM personnel and building occupants of these procedures

15.3.5 External Air Emissions

Air pollution isn't just a threat to health, it also causes damage to the environment. Toxic air pollutants and the chemicals that form acid rain and ground-level ozone can damage trees, crops, wildlife, lakes and other bodies of water. Those pollutants can also harm fish and other aquatic life.

Six common air pollutants are found all over the United States. They are particle pollution, ground-level ozone, carbon monoxide, sulfur oxides, nitrogen oxides, and lead. These pollutants can harm health and the environment, and cause property damage. Of the six pollutants, particle pollution and ground-level ozone are the most widespread health threats.

Sources of emissions to air can be divided between natural and man-made. Natural sources emanate from metabolic products, decomposition, fires, storms and volcanic activity. Man-made sources include fuel combustion in stationary sources (power stations and heating systems) and mobile sources (automobiles), emissions of organic and inorganic pollutants during industrial operations, and losses of organics by evaporation from storage facilities and during use both industrially and domestically.

When a fuel is burnt the principle, products are carbon dioxide, water and heat. Carbon dioxide has been implicated in global warming effects. In addition to potential odor, toxicity and carcinogenic effects, in particular chlorofluorocarbons {CFC' s used as aerosol propellants and refrigerants) have been implicated in the depletion of the ozone layer and associated climatic changes.

Damage to vegetation and buildings by air pollutants is a common occurrence. Particles make buildings and other outdoor structures dirty. Contributions by the multifamily dwellings and mixed-use buildings industry to environmental pollution are partially direct and in relatively small quantities and partially taking place indirectly on a somewhat larger scale, as follows:

- Emissions from burning fossil fuels/gas

  The two types of chemicals that are the main ingredients in forming ground-level ozone are called volatile organic compounds (VOCs) and nitrogen oxides (NO_x). VOCs are released by cars burning gasoline, petroleum refineries, chemical manufacturing plants, and other industrial facilities. The solvents used in paints and other consumer and business products contain VOCs.

  - boilers for the generation of steam, hot water
  - automobiles
  - gas-fired equipment in kitchens, laundries
  - indirectly by consuming energy delivered from power stations
  - power generators (emergency power)
- Emissions from the evaporation of hydrocarbons
  - pesticides (chlorinated hydrocarbons)
- Odors, vapors and mists
  - kitchen and laundry exhausts
  - toilet exhausts
  - paints (especially spray), solvents
- Bacteriological pollutants
  - cooling towers
  - swimming pools
  - waste disposal
- CFC's (chlorofluorocarbons)
  - freon loss from cooling equipment

  - use of spray cans
- Miscellaneous gases:
  - formaldehyde (plywood, chipboard)
  - tri-halogen-methane, chlorine (pools) Methane – a Greenhouse Gas generated as a result of the degradation of vegetable matter
  - Nitrogen Oxides – a range of compounds generated as a result of combustion of fossil fuels which are a precursor to photochemical smog

a) Action for facility management

The major approach is to reduce emissions at source, reduce consumption and to switch to less harmful products, systems, means and technology. The key stages are:

- To identify equipment, materials, processes and operating procedures that contribute to harmful emissions.
- To assess the entire building and its contents with regard to emission hazards and concentration levels
- To develop a realistic action plan

The action plan on air emissions is likely to extend well into the future and will require short, medium and long-term planning.

Short term

- Reduce energy consumption by increasing efficiency and cutting waste.
- Use gas instead of fuel oil, fuel oil instead of electricity.
- Support the reduction of emissions from power stations with the help of available technology to minimize noxious emissions
- Implement operating practices that are not harmful to the environment:
  - Use low Sulphur and higher viscosity oil for fuel burning equipment
  - Carryout regular maintenance for boilers and power generators to ensure optimum combustion
  - Avoid all products manufactured from oil derivatives (plastic bags, etc.)
  - Avoid using pesticides which are non-biodegradable. Use biological methods where feasible
  - Implement procedures for the handling, storage and use of paints
- Avoid all products manufactured by or with CFCs, etc., such as foams, aerosol sprays, and solvents.
- Prepare Inventories of the type, age, condition, expected operating life and energy consumption of all current plant, giving priority to that which contains the largest volume of refrigerant. It should list individually chillers and commercial refrigeration equipment. Include recovery and recycling equipment.

Medium term

- Purchase only efficient equipment and other modern technology items.

- Switch to district heat, e.g. utilizing waste heat from power plants.
- Use co-generation (electricity and use of waste heat).

Long term

- Switch from fossil fuels to renewable energy sources, such as solar, wind, bio-conversion.
- Encourage implementation of hydrogen-gas technology (produces water instead of carbon dioxide).
- When selecting refrigerants for HVAC and cooling equipment, the ozone depletion potential and the global warming potential should be researched and calculated. The lifetime of the equipment, refrigerant charge, and amount of refrigerant are all values that must be considered when selecting refrigerants.

15.3.6 Noise

The acoustic environment in a building can be a crucial factor in its success or failure. At the extreme end of the spectrum of performance requirements are spaces such as concert halls and recording studios that both have exacting standards. Mixed-use buildings are a particularly interesting case as they incorporate a wide range of activities within a single building. Conference and meeting spaces, that have strict requirements of acoustic performance, may be positioned be in close proximity to fitness facilities occupied with noisy activities.

Noise is any kind of sound which people consider undesirable disturbing, annoying, and which can have a number of detrimental effects including damage to health. Noise is as much an environmental issue as water, air and soil. It is generally acoustic waste from man-made equipment, devices or actions. Quiet is the condition in which human beings generally feel well and in which they can relax, recover, rest or concentrate.

A reasonably low sound level throughout the occupied areas of the facility is of paramount importance to its acceptance as an establishment of quality. Special attention should be paid to urban areas as lower wind speeds, the existence of potentially high ambient noise. Of equal importance is the desired degree of privacy provided by a low level of sound transmission between adjoining rooms or spaces. Similarly, staff areas must be controlled, for general well-being and for productivity reasons.

External environmental conditions such as noise, affect internal conditions and therefore can prompt a change in occupant behavior. Thus, relationships between occupants and the internal conditions are bidirectional, as each influences the other.

The intensity of the noise is a function of the sound pressure level most commonly expressed as dBA, which is used to indicate the human response to a noise. The instruments for measuring sound are normally able to report a range of measures of the noise source, including the minimum and maximum and on a time-averaged basis.

a) Objectives

- Elimination/minimization of noise levels to create and maintain a suitable environment by addressing door-step localized issues
- Prevent/minimize adverse psychological, physiological or physical effects to building occupants.

- Prevent annoyance/irritation of third parties (neighbors, tenants).
- Minimize possible revenue loss caused by annoyed customers and visitors, who may decide not to return and use the facilities of competitors.

How to achieve these objectives

- Obtain required local/regional standards for acceptable noise levels for each area/location
- Prepare a summary of known problems (complaints)
- Analyze all problems and possible corrective measures to achieve desirable levels;
- Identify all possible sources of noise in the premises (exterior/interior);
- Conduct an audit: prepare a noise 'map' of each location
- Specify target noise levels for each area and critical location
- Establish a program of compliance by:
  - defining planned methods with detailed descriptions
  - cost analyses including evaluation of alternatives; attenuation expected
  - setting of priorities to achieve the agreed targets stating start and finish time

b) Acceptable Noise Levels

The following sound pressure levels are recommended.

Figure 15.5

Recommended Maximum Background Noise Levels.

Type of Activity	Recommended Ambient Sound Level dB(A)
Board and conference rooms	30-35
General office areas	40-45
Private offices	35-40
Open plan areas	35-40
Libraries - reading areas	40-45
Public circulation	40-45
Restaurants	40-45
Lecture theatres - up to 250 seats	30-35
Lecture theatres - more than 250 seats	25-30
Squash courts	50-55
Residential apartments	30-35
Halls, corridors, lobbies	35-40
Service / support areas	40-45

c) Evaluations

Noise control is a process that will constantly bring improved results as knowledge grows. A regular measurement program of noise levels will enable to evaluate success in managing noise.

15.3.7 Pesticides and Herbicides

Pesticides and herbicides are terms to describe a large group of chemical agents developed to kill unwanted life forms. They have a wide range of targets, composition and associated hazards. They are widely used in building facilities, dependent on local conditions, most often in kitchens, waste storage areas, residences and building grounds. Pesticides and herbicides can cause a range of health problems in humans and other animals, including eye, lung, throat and skin irritation, dermatitis and poisoning, and long-term effects, including cancers and birth defects. They persist in the environment.

Pesticides are used to control insect and other animal infestations, herbicides against 'weeds'. Once applied, the chemicals may take considerable time to break down and become inactive. Additionally, they may become concentrated as they pass up the food chain. Residues may remain in the environment for considerable periods. Environmental problems associated with these agents may therefore persist or appear long after the initial application.

Mixed-use buildings use a variety of chemicals to control pests and weeds. Use will depend on the location of the facility and problems experienced. Building facilities located in the tropical environments may require a greater use of insecticides (for cockroaches, ants and mosquitoes) whereas facilities located in urban locations may have problems with rodents.

a) Concerns

Pesticides, herbicides and other chemicals used to control harmful living organisms are by their nature potentially dangerous to human health and the environment. Short-term exposure to these substances can cause a range of problems, including eye, lung, throat and skin irritation, dermatitis and poisoning. They can also have long-term effects, causing cancer and birth defects. As well as affecting humans, these chemicals can also cause harm to other living organisms, some adverse effects taking years to appear.

In addition, as these chemicals do not break down very quickly in the environment, they have the potential to, build up to high concentrations in the soil and water.

The objective is to safeguard the welfare of all building occupants and visitors and to protect the general environment, by replacing hazardous chemicals with less hazardous alternatives and controlling and minimizing the use of hazardous substances. Pesticides should only be used as a last resort and sparingly. As a general rule, avoid products that are labeled "danger—poison" as those tend to be the most toxic.

b) Integrated Pest Management

There are many ways to prevent and control pests without using toxic chemical pesticides or insecticides. An Integrated Pest Management (IPM) Program provide alternative measures to traditional pest management practices using biological control and biotechnology focusing on safe ways to control and eliminate pests. IPM emphasizes the use of physical barriers, biological controls and other natural forms of pest control to minimize the use of pesticides to the greatest possible degree.

IPM explores the ecological approaches in alternative solutions, such as biological control agents, parasites and predators, pathogenic microorganisms, pheromones and natural products as well as ecological approaches for managing invasive pests, rats, suppression of weeds, safety of pollinators.

c) Alternatives for Safer Lawn Care

Integrated Pest Management or IPM is an approach to pest control that utilizes regular monitoring to determine if and when treatments are needed and employs physical, mechanical, cultural, biological and educational tactics to keep pest numbers low enough to prevent intolerable damage or annoyance. Least-toxic chemical controls are used as a last resort.

IPM methods lead to fewer pesticide applications, reducing contamination of air and water. Generally, the companies on this list emphasize monitoring, exclusion, sanitation, baits, and use least-toxic pesticides as a last resort. Some of the termite companies offer non-toxic treatments such as microwaves, heat, and electrogun. Others offer least-toxic treatments such as borates or termite baits.

15.3.8 Hazardous materials

A variety of hazardous materials or materials generating hazardous waste are used in a building facility's daily operations. They must be handled, stored and disposed of carefully. In safeguarding health and the environment, a hazardous materials program should be implemented to minimize the use of hazardous materials, use more acceptable alternatives, restrict use to trained personnel and ensure that storage, labelling, handling and disposal practices meet recognized standards.

a) Sources of Hazardous Materials

Overall cleaning standards should reduce the exposure of building occupants, and visitors to chemical, biological, and particulate matter that may be harmful to human health, and the built and natural environments, and based on LEED standards.

The facility manager should develop a written facility specific Green Cleaning/Operations Plan that comprehensively describes the methods by which the facility is cleaned effectively while protecting occupant health and the environment. The Green Cleaning Plan should identify use of chemicals according to their sources, which include the following:

- Food and beverage operations
- General Housekeeping operations
- Laundry Operations
- Maintenance and engineering
- Leisure Facilities
- Other facilities not stated above

b) Hazards

The hazards associated with some of the chemicals used in facility operations are described below:

- Bleaches
- Caustic cleaners
- Acids
- Solvents

In order to meet the objectives, there is a need to undertake the following:

- Identify and record where hazardous materials are being used, ascertain what they are being used for and the reasons for their use;
- Assess the hazards associated with their use;
- Identify, where possible, environmentally preferable alternatives
- Review handling, storage, labelling and disposal procedures

In some instances, it might not be possible to use less hazardous materials. Consequently, there will be a need to ensure that the effects of their continued usage and disposal is minimized. To achieve this facility management should have copies of a Hazardous Materials Manual covering the use, handling, storage and disposal of hazardous materials. The manual would need to be in a format which can be readily and regularly updated. Manufacturers and suppliers should be required to provide a Material Safety Data Sheet for every chemical supplied. These pages must be compiled into a fully documented manual covering all hazardous materials

15.3.9 Fuel storage

In the building industry fuel storage is widely used to fire boilers for the production of steam and hot water. Emergency generators, using diesel oil, are provided in many buildings. In suburban areas propane gas is used widely in kitchens. To store fuel, tanks are installed mostly underground, but also above ground, including those for propane gas. Fuel oil storage tank sizes range from small day service tanks of 50-gallon capacity to large multiple tanks of up to 13,250 gallons each.

When fuel oil is released uncontrolled, the fuel storage facility would pose environmental and social risks:

- It coats the soil and seeps down on top of the water table and moves with the ground water, even over great distances
- As a liquid it can readily cause fires and as vapor it fans out, leading to possible explosion hazards.

All environmental risks can be minimized and managed through implementing preventative measures and sound management systems. The Department of Environmental Protection in most states require Third Party Underground Storage Tank (UST) Inspections to be conducted, document the results of their on-site inspections; review facility record keeping to ensure that it meets UST Program requirements; and report their findings. It is a requirement to complete these inspections to permit new facilities as well as to ensure compliance for existing tank storage facilities.

To support sound decision making that is consistent with the principles of sustainable development, the consideration of environmental effects must begin early in the conceptual planning stages of the project, before irreversible decisions are made. In this way, environmental assessment can support the analysis of options and identify issues that may require further consideration.

a) The environmental assessment

- **Scope and nature of potential effects**. The analysis should build on a preliminary scan to describe, in appropriate detail, the scope and nature of environmental effects that could arise from implementing the proposed project. Environmental effects, including cumulative effects, could result from the use of, or changes in, atmospheric, terrestrial or aquatic resources, physical features or conditions. The analysis should identify positive as well as adverse environmental effects.

- **The need for mitigation**: The assessment should consider the need for mitigation measures that could reduce or eliminate potential adverse environmental consequences of the proposed project. Mitigation could include, for example, changes in the project parameters, conditions that may need to be placed on specific components of the project or activities arising from the project, or compensation measures.
- **Scope and nature of residual effects:** The analysis should describe, in appropriate detail, the potential environmental effects that may remain, taking into account mitigation measures.
- **Follow-up:** The environmental assessment should also consider the need for follow-up measures to monitor environmental effects of the policy, plan or program, or to ensure that implementation of the proposal supports the department's sustainable development goals.
- **Public and stakeholder concerns:** The analysis should identify for decision makers, where appropriate, concerns about the environmental effects among those likely to be most affected, and among other stakeholders and members of the public.

b) Site Analysis

Tank installation is a site-specific process which is directly influenced by local physical conditions. A thorough knowledge of the site characteristics is therefore critical. Examples of site characteristics would include but not limited to:

- Location of buildings
- Location of underground structures (e.g. basements, underground cables, utility trenches, etc.)
- Soil structure and permeability
- Depth to the water table
- Proximity to sensitive areas (flora, fauna, residential areas vs commercial areas)
- Proximity of surface waters
- Acidity of Soil (pH)
- Historic Contamination

c) Emergency Planning and Response

Contingency planning is one of the most important action elements in response to "release" or "spill" incidences. Proper planning for effective initial response to a release and/or spill incident is critical to limiting both the degree of hazard and the extent of contamination. Contingency planning is essential for a successful response to a hazardous material incident.

An emergency plan should be comprehensive, yet clear and concise for easy reference during an Emergency. The roles and responsibilities of all players should be clearly defined, areas of risk and hazard should be predetermined, and procedures for dealing with a release should be clearly outlined.

15.4 Operations and Maintenance

The primary objective of an organization's maintenance function is to maximize the availability and efficiency of the organization's assets throughout their expected lifecycles at the lowest possible cost. An overall proactive maintenance strategy that extends the life of assets and improves overall efficiency and reduces maintenance costs is essential. Maintenance management dictates that "what you do not maintain today you will have to maintain tomorrow, except that the costs will be far greater".

A Predictive Maintenance strategy relies on the use of technology and tools to constantly monitor an equipment's efficiency and wear to provide advanced warning of an impending failure or loss of function. The advantage of this methodology is that this monitoring occurs while the asset is running. The next generation of wireless protocols is going to change the way AR and VR are being converged with the Internet of Things (IoT) into mixed reality (MR) to provide a more seamless and realistic experience.

It is important to prepare for the future by investing in high-quality facilities management software skilled at creating automated work orders, forecasting occupancy rates etc. Technology in and of itself doesn't change anything for better or worse. People do. Technology is merely an enabler for improved human performance. As we bring in the next generation of automation workers, they can utilize modern technology to bring new insights and innovations to operation and maintenance processes. Workers with domain experience are key in taking advantage of technology shifts and applying it to operations.

New and improved technology is, by definition, always available. It takes a facility management team with the foresight and planning to find out how best to leverage the right technology for them, and to employ that technology to help them win against the competition. That's where a good platform-independent automation solutions provider can help analyze the current processes and sort through the myriad of options to help each individual company modernize, save, and jump forward in the markets they serve.

Improvements and developments in the latest technologies such as AI,AR and VR will allow facility managers to fully automate the most complex plant and equipment systems, and predictive algorithms will provide personnel with the information needed to fix issues before equipment breakdowns or safety issues occur.

Facility Managers that embrace the digital enterprise will gain a competitive edge. These facility operators will have a digital twin of their entire value chain, enabling them to perform simulation, test, and optimization in a completely virtual world. And they can resolve problems and optimize their enterprise in the virtual world before ever committing any physical resources in the real world. The ability to collect vast amounts of data in near real-time from a broad range of intelligent connected devices is the foundation of IoT. This data can then be accessed directly, or via the cloud; and unique value propositions could be created through the application of complex analytics and data techniques. In this way, the IoT can, and will, be used to provide unique value propositions and create complex information systems which are greater than the sum of the individual components.

15.4.1 Artificial Intelligence (AI) and maintenance

With the increasing complexity and globalization of supply chains and the advancement of building and workplace automation, there are increasing needs for minimizing downtime with regard to building facility plant and equipment. With the declining number of skilled engineers and the demands for optimization and greater sophistication of repair services for products utilizing digital technologies, more facility management companies are using an artificial intelligence (AI) platform that will integrate real-time operational data in plant and equipment and use the data to determine the equipment service needs in real time.

Most of the data needed to implement a maintenance monitoring system is gathered from the equipment. The maintenance teams should be trained to Through a collaborative approach and working with maintenance teams build awareness about the available data and emphasize the need to unlock the potential of this data to address, define and adopt digital maintenance strategies and building analytical models.

AI is an element of computer software that can make (or enable) intelligent decisions and problem solving through deduction and reasoning. This is made possible by collecting and analyzing vast repositories of relevant data that can inform decisions and provide insights – the fundamental requirements for any form of intelligence.

AI will play an important role in many aspects, from building on cumulative knowledge of historical maintenance strategies and repair actions, to assessing risks in plan and optimization decisions. AI will assist the analytical model to learn continuously, which will increase the accuracy of prescriptions or recommendations based on relationships among various data. This includes data on machinery breakdowns and details of repairs and results of work performed. The facility management team could use the recommendations provided by AI-based advanced digital solutions to increase the speed and accuracy of decision making process.

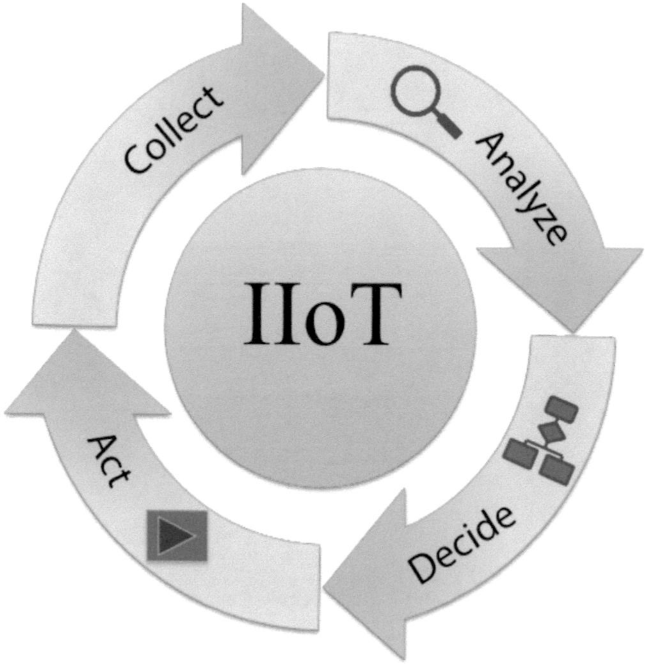

Figure 15.5 Industrial Internet of Things (IIOT)

An important benefit of this might be in being able to identify and eliminate routine and repetitive tasks that can be easily automated. Facility Managers who use service automation already know about the benefits of automating tasks such as recurring planned maintenance and required inspections of critical equipment.

It is a proven fact that AI-powered systems will not require workers with a different skill set. The existing technical skills of workers should only require enhancement through training to get acquainted with the systems, their features and usage. The systems will provide repair recommendations utilizing machine learning, based on repair history, that can work as a knowledge repository specific to a piece of equipment or set of assets. This knowledge repository will provide quick reference for any technician. The systems should be interactive and allow technicians to communicate with a remote expert to analyze issues and provide remedies. This information should then be uploaded into a knowledge repository for AI-based systems to learn from and refer to the next time it is needed. AI based maintenance systems learns continuously and is an ever-evolving system in that respect.

15.4.2 Augmented Reality (AR) and Virtual Reality (VR) in maintenance

Augmented reality (AR) and virtual reality (VR) empowers facility management teams to gain insights into their equipment health. This leads to operational efficiencies, which in turn enhances product quality. These technologies leverage sensors, cameras, smart devices and wearables, and other Industrial Internet of Things (IIoT) tools.

Training becomes easier, technicians are given a visual, hands-on experience in front of the machines leading to maintenance teams leveraging AR-overlaid displays to view the machine's condition, facilitating problem detection ahead of solving it in person. An AR-headset uses technology to guide a technician with instructions on the line of sight.

Remote assistance using AR and VR solutions will enable people in different geographies to connect and troubleshoot problems together. A technical issue in the United States could be resolved by collaborating with an engineer in India using IoT- and voice-enabled AR glasses, thus cutting travel costs and expediting the problem-solving process.

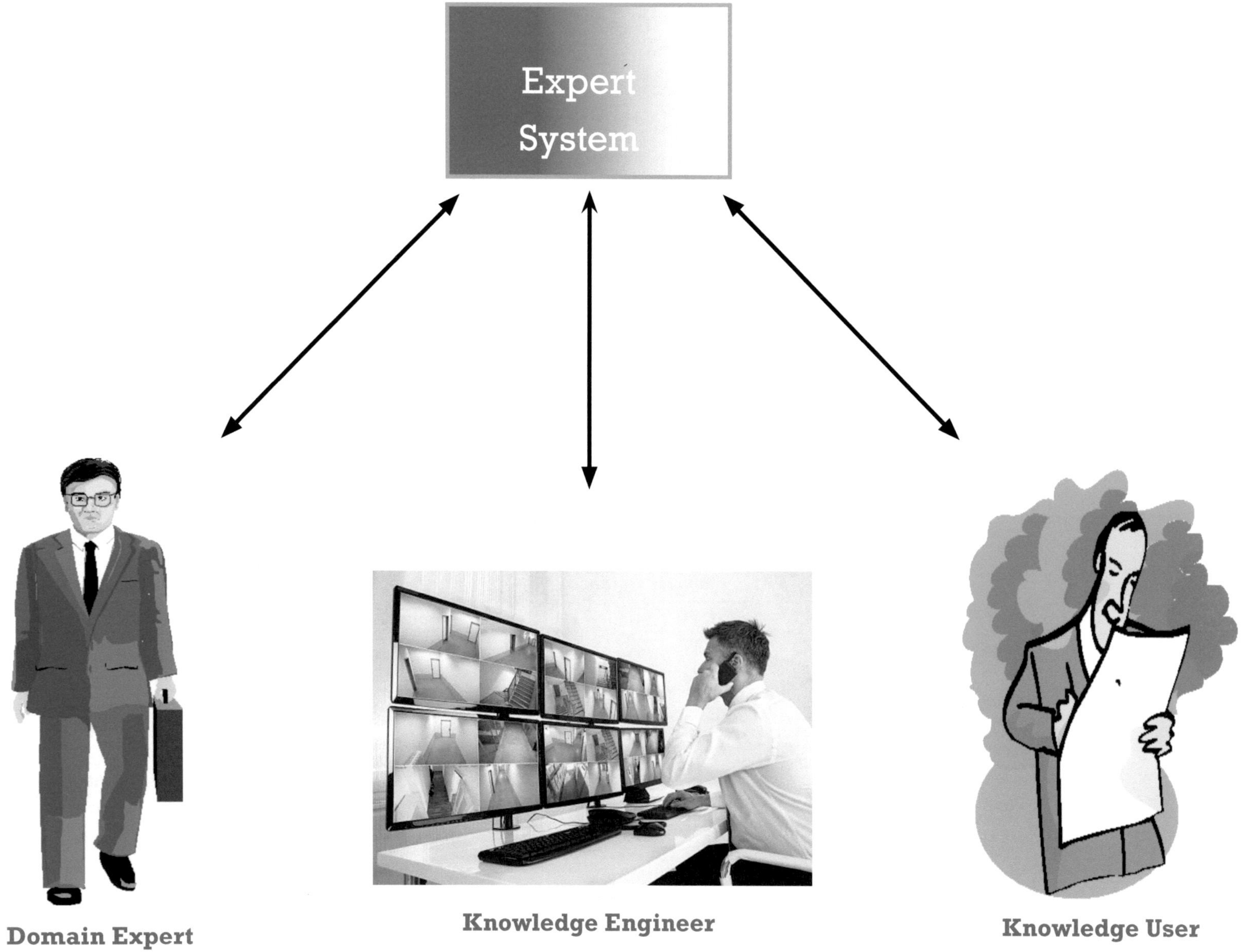
Expert System
Domain Expert
Knowledge Engineer
Knowledge User

Acronym	Definition
ASTM	American Society for Testing and Materials
ACI	American Concrete Institute
AI	Artificial Intelligence
ACMV	Air Conditioning and Mechanical Ventilation
ADA	Americans with Disabilities Act
ADAG	Americans with Disabilities Act Guidelines
AMI	Advanced Metering Infrastructure
AMR	Automatic Meter Reading
ANSI	American National Standards Institute
AHU	Air Handling Unit
AHRI	Air Conditioning Heating and Refrigeration Institute
AC	Air Changes
APA	American Plywood Association
AV	Audio Visual
AR	Augmented Reality
ASHRAE	American Society of Heating, Refrigerating, and Air-Conditioning Engineers
BMS	Building Management System
BIM	Building Information Modelling
BIPV	Building-Integrated Photovoltaic
BREAM	Building Research Establishment Assessment Method
BS	British Standards
CMU	Concrete Masonry Units
C	Commissioning
CRI	Carpet Rug Institute
CPU	Central Processing Unit
CFC	Chlorofluorocarbon.
CAPEX	Caital Expenditure
CCHP	Combined Cooling, Heat and Power
CHP	Comined Heat and Power
CEIMP	Construction Environmental Impact Management Plan

Acronym	Definition
Cp	Pressure Coefficient
CO_2	Carbon Dioxide
CFD	Computational Fluid Dynamics
CRI	Color Rendering Index
CIBSE	Chartered Institution of Building Services Engineers
DALI	Digital Addressable Lighting
DBB	Design Bid Build
DID	Design Intent Document
dB	Decibel
dBA	A-weighted Decibels
DCS	District Cooling System
DCV	Demand Controlled Ventilation
DOAS	Dedicated Outside Air System
DOE	Department of Energy
DMX	Digital Multiplex Signal
EIA	Environmental Impact Assessment
EIS	Energy Information System
EDC	Environmental Design Consultant
EfW	Energy from Waste
EMAS	European Eco-Management and Audit Scheme
EMS	Energy Management System
EMO	Energy Management Opportunities
EPD	Environmental Product Decleration
EQ	Energy Quotient
ERV	Energy Recovery Ventilator
ETS	Environmental Tobacco Smoke
EPR	Extended Producer Resposibility
EPA	Environmental Protection Agency
FSC	Forest Stewardship Council
GHSP	Ground Source Heat Pump
GHG	Green House Gas

Acronym	Definition
GFI	Ground Fault Interrupter
GFCI	Ground Fault Circuit Interrupter
Gbits/sec	Gigabitsper second
GSM	Grams Per Square Meter
HVAC	Heating Ventilation and Air Conditioning
HVAC & R	Heating Ventilation Air Conditioning and Refrigeration
HRV	Heat Recovery Ventilator
HSIA	High Speed Internet Access
HCFC	Hydrochlorofluorocarbon
IaaS	Infrastructure as a Service
IBC	International Building Code
IPC	International Plumbing Code
IAQ	Indoor Air Quality
IgCC	International Green Construction Code
ICT	Information and Communications Technology
IECC	International Energy Conservation Code
IES	Illuminating Engineering Society
ICC	International Code Council
IIC	Impact Insulation Class
IIOT	Industrial Internet of Things
ICF	Insulated Concrete Forms
IESNA	Illuminating Engineering Society of North America
IMC	International Mechanical Code
IEC	InternationalElectrotechnical Commission
IP	Internet Protocol
IR	Infra Red
IT	Information Technology
IBC	International Building Code
IRC	International Residential Code
IOT	Internet of Things
IMCC	Intelligent Motor Control Center

Acronym	Definition
JIT	Just in Time
KW	Kilo Watts
Kwh	Kilo Watt Hour
LAN	Local Area Network
LEED	Leadership in Energy and Environmental Design
LEED-NC	LEED New Construction
LED	Light Emitting Diode
LVT	Luxury Vinyl Tile
LVP	Luxury Vinyl Plank
LV	Low Voltage
LEED AP	Leadership in Energy and Environmental Design -Accredited Professional
LSZH	Low-Smoke-Zero-Halogen
L_{Aeq}	A-weighted Equivalent Sound Pressure Level in dB
LCA	Life Cycle Assessment
MATV	Master Antenna Television
MCC	Motor Control Center
MDM	Meter Data Management
MEP	Mechanical Electrical Plumbing
MDF	Medium Density Fiberboard
MV	Medium Voltage
NAC	Network Access Control
NFPA	National Fire Protection Association
NSPI	National Spa and Pool Institute
NRC	Noise Reduction Coefficient
NEC	National Electrical Code
OSHA	Occupational Safety and Health Administration
OITC	Outside-Indoor Transmission Class
OPEX	Operational Expenditure
OPR	Owner's Project Requirements
PoE	Power over Ethernet

Acronym	Definition
PaaS	Platform as a Service
PTT	Public Telephone & Telegraph
PICP	Permeable Interlocking Concrete Pavements
PVs	Photovoltaics
PLC	Programable Logic Controller
PMAC	Permanent Magnetic Alternating Current
PVC	Polyvinyl Chloride
PBX	Private Branch Exchange
PC	Personal Computer
PCB	Polychlorinated Biphenyls
PIR	Passive Infra-Red
PMS	Property Management System
QOS	Quality of Service
RFP	Request for Proposal
RF	Radio Frequency
SaaS	Software as a Service
SBSDG	Sustainable Building Standards and Guidelines
SPL	Sound Pressure Level
SHGC	Solar Heat Gain Coefficient
STC	Sound Transmission Class
SRI	Solar Reflectance Index
SFI	Sustainable Forestry Initiative
SC	Shading Coefficient
SMPS	Switch-Mode Power Supplies
SSH	Secure Shell
SCAQDM	South Coast Air Quality Management District
TBL	Triple Bottom Line
TV	Television
UHI	Urban Heat Islan
UV	Ultra Violet
UPC	Uniform Plumbing Code

Acronym	Definition
UPS	Uninterrupted Power Supply
UL	Underwriters Laboratories
USGBC	U.S. Green BuildingCouncil
UTP	Unshielded Twisted Pair
VGA	Video Graphics Array
VOC	Volatile Organic Compounds
VFD	Variable Frequency Drive
VRF	Variable Refrigerant Flow
VRV	Variable Refrigerant Volume
VR	Virtual Reality
VAV	Variable Air Volume
VLT	Visual Light Transmission
VLAN	Virtual Local Area Network
VPN	Virtual Private Network
WAN	Wide Area Network
WiFi	Wireless Networking Technology 802.11 and above (Wi-Fi Alliance)
WAN	Wide Area Network
WWAN	Wireless Wide-Area Network
WAP	Wireless Access Point
WEP	Wired Equivalent Privacy
WLAN	Wireless Local Area Network
WMP	Waste Management Plan
WHO	World Health Organization
WPA	Wi-Fi Protected Access
WC	Water Closet

Technical Services CONSULTANTS

Printed in the United States
By Bookmasters